Towards Future Technologies for Business Ecosystem Innovation

RIVER PUBLISHERS SERIES IN COMMUNICATIONS

Series Editors

ABBAS JAMALIPOUR
The University of Sydney
Australia

JUNSHAN ZHANG
Arizona State University
USA

MARINA RUGGIERI
University of Rome Tor Vergata
Italy

Indexing: All books published in this series are submitted to the Web of Science Book Citation Index (BkCI), to CrossRef and to Google Scholar.

The "River Publishers Series in Communications" is a series of comprehensive academic and professional books which focus on communication and network systems. Topics range from the theory and use of systems involving all terminals, computers, and information processors to wired and wireless networks and network layouts, protocols, architectures, and implementations. Also covered are developments stemming from new market demands in systems, products, and technologies such as personal communications services, multimedia systems, enterprise networks, and optical communications.

The series includes research monographs, edited volumes, handbooks and textbooks, providing professionals, researchers, educators, and advanced students in the field with an invaluable insight into the latest research and developments.

For a list of other books in this series, visit www.riverpublishers.com

Towards Future Technologies for Business Ecosystem Innovation

Editors

Ramjee Prasad

GISFI, India and President
CTIF Global Capsule
Aarhus University
Denmark

Leo P. Ligthart

Chairman, CONASENSE
The Netherlands

Published, sold and distributed by:
River Publishers
Alsbjergvej 10
9260 Gistrup
Denmark

River Publishers
Lange Geer 44
2611 PW Delft
The Netherlands

Tel.: +45369953197
www.riverpublishers.com

ISBN: 978-87-93609-77-8 (Hardback)
      978-87-99923-70-0 (Ebook)

Contents

**3 5G and Telemedicine: A Business Ecosystem Relationship
    within CONASENSE Paradigm   41**

Ambuj Kumar, Sadia Anwar and Ramjee Prasad

Preface

Integration of communication, navigation, sensing and services (CONASENSE) is one of the major research areas under the umbrella of CTIF Global Capsule (CGC). CGC is a non-profit association that has an ambition of leading the convergence of business, economics and technology, to provide benefit to the society. CGC aims at embracing the aspirant to aggregate and collaborate in striding for the new scientific paths.

Present book, Towards future technologies for business ecosystem innovation, is the 6th volume in the series of CONASENSE and consists of 9 chapters.

Chapter 1 presents about digitization and ubiquitous connectivity that demands more capacity, quick response, lower delays and reliable communication with good quality of service and better user quality of experience. Usable security and privacy are of the prime importance while designing future wireless ecosystem (FWE).

Chapter 2 brings up the vision that the Access Network (AN) in smart city (SC) will develop towards a flexible and cognitive Unified Wireless Access (UWA) infrastructure with intelligence spread down to the level of each single user.

Chapter 3 surveys the current state of telemedicine along with examining the characteristics of 5G technology.

Chapter 4 makes an overview of the most interesting business applications in which these systems could play an important role. Then, an overview of traditional and more novel approaches is provided. Finally, the Chapter discusses on the opportunity to bring location/context-based services, and the associated huge market opportunities, as protagonists in 5G systems.

Chapter 5 describes the role of Satellites and Unmanned Air Vehicles (UAVs) in implementing the integrated vision of CONASENSE. It lists out various CONASENSE applications enabled by Satellites and UAVs.

Chapter 6 describes how and why coalitions of sensor nodes form, using the trade-off between the advantage of cooperation (in terms of better performance) and the costs of cooperation (in terms of bandwidth, transmitting information), using the topology of the network. A general game-theoretic framework for wireless sensor network (WSN) is presented, and illustrated by means of an example.

Chapter 7 presents, a practical solution for the automatic maintenance of utility-scale photovoltaic power plants by employing Remotely Piloted Aircraft System (RPAS).

Chapter 8 focuses on the service of CONASENSE and summarized the current situation of network neutrality (NN) in service innovation era.

Chapter 9 reports on studies carried out on Virtual Business Models. Different Virtual Business Model cases are presented.

We are sure that this book will provide the role of CONASENSE for innovating future technologies.

Ramjee Prasad
GISFI, India
CTIF Global Capsule
Aarhus University, Denmark

Leo P. Ligthart
Chairman, CONASENSE
The Netherlands

List of Figures

List of Tables

1

The Enablers of Future Wireless Ecosystem

Vandana Rohokale[1] and Ramjee Prasad[1]

[1]Sinhgad Institutes, Pune, India
[2]CGC, Aarhus University, Herning, Denmark

Abstract

An ecosystem is a strong network of living things such as plants, animals, organisms and non-living things like weather, sun, earth, soil, climate, atmosphere which keeps whole earth system healthy. In similar way, wireless ecosystem demands a win-win convergence of all the cyber physical systems with existing wired and wireless networks with green energy flow among various life cycles in a cost-efficient and user friendly manner. Wireless communication was earlier considered to be a luxury. But now it is becoming the necessity of the human life. In the age of forthcoming Internet of things (IoT), machine to machine communication has to play a vital role. Digitization and ubiquitous connectivity demands more capacity, quick response, lower delays and reliable communication with good quality of service and better user quality of experience. Usable security and privacy are of the prime importance while designing future wireless ecosystem (FWE).

Keywords: Internet of Things, Machine to Machine Communication, Future Wireless Ecosystem, Convergence, Cyber Physical System.

1.1 Introduction

Imagine a situation as depicted in Figure 1.1 below where a working woman reaches home at evening tired and mentally exhausted due to whole day office work. Then she seats in a chair for a while worrying about the duties she needs to perform at home front like cooking for dinner, dish washing,

Towards Future Technologies for Business Ecosystem Innovation, 1–20.

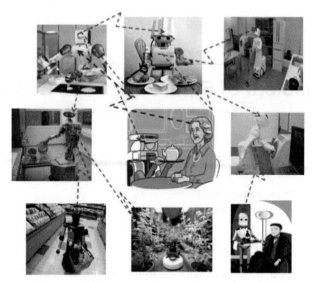

Figure 1.1   Cooperative Machine to Machine Communication.

clothes washing, gardening, home cleaning, purchasing some goods from supermarket, please her husband when he returns home etc. But what if she suddenly gets a ready cup of tea from a robot and the work list she was thinking of has been sensed by a nearby robot and it has communicated these messages to respective robot machines and all of them have started performing their tasks perfectly. And to her surprise,within half an hour everything is perfectly done! How relax she will feel and then she can utilize her energies in revitalizing herself and will be able to pay proper attention to her family members.

The day is not far away when we are going to witness these kind of scenarios around us when machine to machine communication, augmented reality and cooperative communication will be integrated altogether. Then the machines will be able to sense, command and control the situations with proper decision making capability. Machine to machine cooperative communication with these intelligent capabilities like sense, command and control other machines will help in bringing converged wired and wireless networking services needed for fifth generation of mobile communication into reality.

Currently, wireless communication and mobile computing are the buzz words for the telecom industry. For multimedia applications, the user needs higher data rates at which data transactions can take place efficiently.

Gigabit wireless communication is the dream which is being chased by scientists and researchers. Different wireless networking technologies include cellular networks, Wi-Fi Networks, WiMAX Networks, Wireless sensor networks etc., which can deploy cooperation for numerous benefits of cooperation. Cooperation strategy can be effectively applied for these well-known networking techniques [1].

For bringing this kind of ecosystem into existence, the world needs a robust wireless network with high capacity and low latency which should be optimized for scalability, reliability, latency, throughput, security, privacy and energy efficiency. The chapter is organized as follows. Section 1.2 gives detailed information about the evolution of wireless communication networks from first generation to fifth generation. Section 1.3 throws light on various enablers of the future wireless ecosystem such as Internet of Things (IoT) and machine to machine communication, Visible Light Communication, introduction to intelligence in the wireless networks etc. Segmentation strategy for future wireless ecosystem is elaborated in Section 1.4. Usable security, privacy and its need is highlighted in Section 1.5. Section 1.6 contains some discussions and summary.

1.2 Evolution of Wireless Communication

Alexander Graham Bell in 1870s designed a device which could transmit speech on electric lines and that is known as telephone. He extended his experiments with telephone to the concept of harmonic telegraph in which he proved that several notes can be transmitted simultaneously along with the single wire if notes differ in pitch [2]. Bell also invented Photophone around 1880s, which could transmit sound over a light beam [3]. During 1960s, Kao and Hockham proposed optical fibers made up of glass for the transmission of speech signals. Optical fibers are tiny strands of ultra-pure glass which can transmit voice, data, text and images in digital format including thousands of telephone calls on light photons.

Michael Faraday, James Clerk Maxwell and Heinrich Hertz laid the foundation for wireless communication by proposing theory of electromagnetism with introduction of different mechanisms like Faraday's law of Electromagnetic Induction, Maxwell's equation and Electromagnetic Radiation Photoelectric Effect. Guglielmo Marconi extended theoretical work by these scientists with the help of laboratory experiments to develop Wireless Telegraphy. Marconi made use of electromagnetic waves to transmit signals over large distances and he acquired patent for Radio in 1904 [4]. Bell system (AT&T) proposed a Broadband Urban Mobile System in early 1947, requesting 40 MHz band for its implementation to FCC.

Mobile Radio Telephone system emerged as per cellular system which was analog in nature with half duplex communication capabilities. It was also considered as zeroth generation of mobile communication. It comprises of several techniques such as Advances Mobile Telephone System (AMTS), Mobile Telephone System (MTS), mobile Telephone System D (MTD), Public Land Mobile Telephony (PLMT), Push to Talk (PTT) and Improved Mobile Telephone Service (IMTS). Initially, these mobile phones were placed in vehicles such as truck, cars, etc. with coverage of 20 Kms. Only 25 channels were available so this technique could not spread into public. It was restricted only for limited business applications.

1.2.1 First Generation (1G) of Mobile Communication

With the evolution of Cellular technology, the first generation of mobile communication was initiated in 1980s. The 1G mobile phones were analog in nature with frequency division multiple access (FDMA) technique employed for spectrum access. The demand for radio spectrum in 1G was large due to FDMA technique. Japan's Nippon Telegraph and Telephone (NMT) deployed analog mobile phones in Nordic European countries like Finland, Sweden, Denmark and Norway in 1981. Two versions of NMT were introduced with frequency bands of 450 MHz and 900 MHz with respective names given as NMT-450 and NMT-900.

Advanced Mobile Phone System (AMPS) was the first analog cellular system deployed in North America. In 1983, it was commercially deployed in America, in Israel in 1986 and in Australia in 1987. AMPS were the real cellular system which was accepted by mass user market but it was not secure due to lack of encryption. It was prone to eavesdropping attack via a scanner and was also vulnerable to cell phone cloning [5]. As depicted in Figure 1.1, 1G mobile system consisted of many standards including Advance Mobile Phone Service (AMPS), Nordic Mobile Telephony (NMT), and Total Access Communication System (TACS) etc. Also, the user equipment was very much bulky with bigger size and weight as shown in Figure 1.2. First generation mobile phone could transmit only analog voice with data rate ranging from 2 kbps to 14.1 kbps.

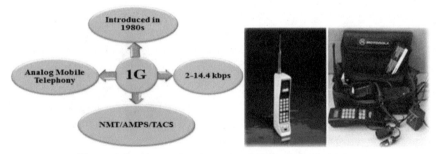

Figure 1.2 First Generation (1G) Cellular Communication.

1.2.2 Second Generation (2G) Cellular Communication

Second generation (2G) Digital Mobile Phone System emerged around 1990s. Two competent standards were developed named Global System for Mobile (GSM) and Code Division Multiple Access (CDMA). GSM was introduced by Europe and United States developed CDMA. GSM network was firstly launched in Finland during 1991. 2G mobile phones introduced the ability to access media content through Internet on mobile phone (MMS). 2G digital mobile phones were able to provide data rates ranging from 64 kbps to 144 kbps. Second generation mobile System overview is shown in Figure 1.3. GSM, TDMA (IS-136) and CDMAone(IS-95A) were the standards evolved for 2G. The first mobile equipped with full internet service were launched by NTTDoCoMo during 1999 in Japan. 2.5G mobile system is the interface between 2G and 3G. 2.5G provides additional features such as

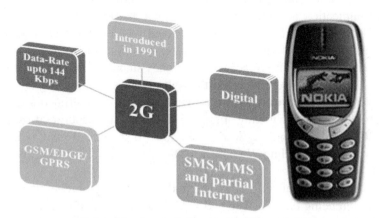

Figure 1.3 Standards Provided by 2G and 2.5G.

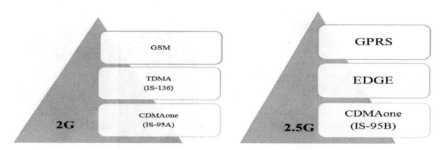

Figure 1.4   Standards Provided by 2G and 2.5G.

packet switched connection (GPRS) and 2.75G provides enhanced data rate (EDGE). GPRS comes under the umbrella of GSM whereas EDGE is used by both GSM and CDMAone. 2.5G networks provide much higher data rates ranging from 40 kbps to 236 kbps and 2.75G offers data rates up to 384 kbps. Figure 1.4 indicates the standards provided by 2G and 2.5G mobile services. The modulation technique used by second generation mobile communication is GMSK. This generation provides services such as multimedia messaging and internet access.

1.2.3 Third Generation (3G) of Mobile Communication

Around 1980s International Telecommunication Union (ITU) has set a standard for 3G called International Mobile Telecommunication-2000 (IMT-2000). Due to the seamless connectivity feature provided by 3G mobile phones, the world really got shrink into a small connected village. 3G mobile telecommunication provides applications such as global roaming, wireless voice telephony, mobile internet access, fixed wireless internet, video telephony and mobile TV. 3G technology supports data rates from 144 kbps to 2 Mbps. It uses multiple access techniques such as TDMA, FDMA and CDMA. Voice calls are connected through circuit switching but the multimedia data are sent through packet switching but the multimedia data are sent through packet switching. Figure 1.5 depicts the superficial overview of the 3G mobile technology whereas Figure 1.6 shows different variants of 3G or IMT-2000 including W-CDMA, CDMA2000, UMTS, DECT, WiMAX etc. [6].

3G supports enhanced audio and video streaming, higher data rates, video conferencing facility, Web and WAP browsing with high speeds, IPTV support and global roaming. 3.5G service is known as High-Speed Downlink

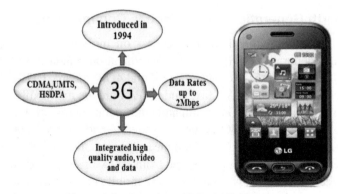

Figure 1.5 Overview of 3G Mobile System.

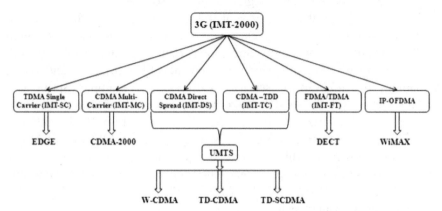

Figure 1.6 Standards Provided by Third Generation Mobile Communication.

Packet Access (HSDPA) mobile telephony protocol. It is an enhanced version of 3G, popularly known as 3G UMTS which is packet based service supporting higher data transfer speeds. HSDPA in WCDMA downlink provides data transmission speeds of 8–10 Mbps for regular systems and 20 Mbps for MIMO systems. Adaptive modulation techniques are used for achieving such higher data rates [7]. 3.75G mobile telephony system is called High Speed Uplink Packet Access (HSUPA) which is basically UMTS-WCDMA uplink technique. 3.9G Mobile system is the first step towards Long Term Evolution (LTE) with Single Carrier Frequency Division Multiple Access (SCFDMA) which is truly digital wideband packet data transmission system and is considered to be the dawn for the birth of 4G.

1.2.4 Fourth Generation (4G) of Mobile Communication

Wireless Digital Broadband Internet access on mobile phones is possible due to fourth generation (4G) mobile communication standard. In December 2010, ITU approved IMT-Advanced as official 4G standard. LTE-Advanced, Wireless MAN-Advanced. WiMax2 and HSPA+ are the main building blocks for 4G. It was difficult for 3G to roam and interoperate across global networks due to multiple standards. 4G is the integrated versions of Wireless LAN's, Bluetooth, Cellular Networks etc. with the beginning of convergence of wired and wireless networks. Whooping speed from 100 Mbps to 1 Gbps is the most attractive feature of 4G. It is considered as a mobile multimedia with anytime anywhere connectivity and global mobility support and customized personal service network system with integrated wireless solution.

Figure 1.7 gives the overview of 4G system. Some of the important features of 4G include all IP packet switched network, high data rates upto 1 Gbps, seamless connectivity and global roaming, Interoperability with existing standards. Smooth handovers and high QoS [8]. The underlying technology for different 4G services is diverse. For example, Sprint uses WiMax technology for its 4G network with provision of data rates up to 10 Mbps whereas Verizon wireless makes use of Long Term Evolution(LTE) 4G technology and can provide speeds ranging from 5 Mbps to 12 Mbps. 4G is all IP network in which converged voice and data are transmitted over Internet Protocol. It is fully packet switched network. It uses various multiple access techniques such as OFDMA, CDMA, WCDMA, LAS-CDMA etc. for achieving higher data rates up to 1 Gbps.

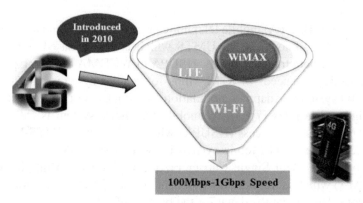

Figure 1.7   Overview of 4G Mobile Communication.

The authors in [9] have visualized the shrinked world with 4G as Global Information Multimedia Communication Village (GIMCV) for which the cellular sales are moving from global to pico-cellular size. Necessity of collaborative functionality of WLANs and WPANs in coordination with 3G systems is justified with GIMCV paving the way for 4G. Key issues in 4G networks are tightly addressed in the book of advanced wireless networks [10, 11]. It throws light on cognitive, cooperative and opportunistic approach which is capable to improve network efficiency further. Issues related to almost all emerging networks like wireless internet, mobile cellular, WLAN, WSN, adhoc networks, bio-inspired networks, active and cognitive networks with cooperative strategies which are integrated in the 4G networks are discussed effectively. Cross layer optimization techniques are well discussed with network information theory by considering adaptability and reconfigurability of the evolving networks.

1.2.5 Fifth Generation (5G) of Mobile Communication

Majority of economic growth of today's wireless world is driven by Information and Communication Technology (ICT). Society and digital economy are closely interconnected with each other via many critical infrastructures like energy and electricity, water, food, transportation, public health, telecommunication and ICT, different government services etc. Previous generation services are either service oriented or operator centric. But the 5G revolution is going to be fully user centric with better quality of experience (QoE) for user. 2G digital systems could effectively replace the fully analog 1G service.

With arrival of 3G network services, 2G went little bit back on foot but still it is actively in function with some important GSMs services. 4G LTE and LTE-Advanced systems are designed especially for the high bandwidth data and videos with higher data rates. But 5G is not going to be the efficient replacement of 4G and previous generation systems. Instead it will be global convergence of almost all advance wireless networking techniques with the concepts like cooperation and cognition. 5G will be like a joint family works in cooperation mode with others through cognition capabilities then 5G with infinite capacity, thousand times higher data rates, delay less than one millisecond along with larger amount of throughputs is going to come to reality. The mobile communication evolution form first generation to fifth generation is shown in Figure 1.8 which depicts gradual improvement in the mobile communication experience.

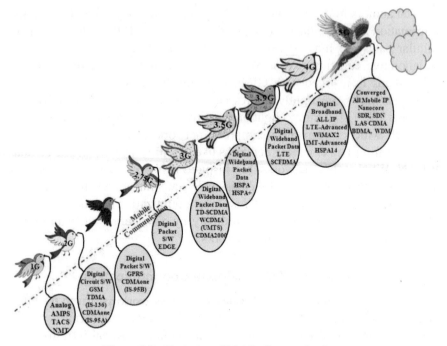

Figure 1.8   Evolution of Mobile Communication

The journey of mobile communication from 1G to 5G is wonderful. But every generation suffered from number of security threats. Analog 1G system faced fraud and eavesdropping kind of attacks. Second generation underwent attacks on digital encryption system. 3G mobile communication has encountered attacks on International Mobile Subscriber Identity (IMSI) and hacking of incoming and outgoing calls. Internet IP based security threats are posing problems in front of 4G. Upcoming 5G may suffer from security and privacy related threats based on converged network and services. The evolution of various threats is depicted in Figure 1.9 along with every generation of mobile communication.

1.3 FWE Enablers

FWE development is dependent on many existing wired, wireless and next generation technologies. The main enablers of FWE include Internet of Things (IoT) and Machine to Machine Communication (M2M), Visible Light Communication (VLC), Intelligent Transportation System (ITS), Artificial

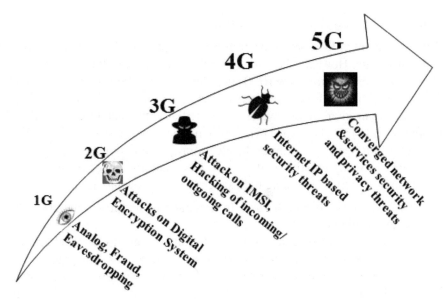

Figure 1.9 Attack evolution along with the Wireless Network Progression.

Intelligence (AI), secure and robust wireless networks, etc. This section throws light on the importance of all these enabling technologies in the development of FWE.

1.3.1 IoT and M2M

Internet of Things (IoT) is the essential convergence of Machine to machine communications (M2M) communications and data analytics. M2M aims for the automation and streamlining of industrial systems and processes whereas IoT has complete transformation capability providing industries with renovation of their products and services at the same time bringing each and every entity into the global internetworking. Service centric M2M is now converging towards user centric IoT [12]. According to Gartner's prediction, around 25 billion devices will be connected globally by 2020 including smart phones, personal gadgets, wearables, connected vehicles, etc. The wireless information transfer among these connected objects should not negatively impact on the sustainability ecological safety of the connected world. FWE should responsibly preserve and protect the environmental, industrial, agricultural and cyber physical system's sustainability, integrity and safety. IoT evolution is being successfully utilized for solving the honeybee crisis all over the world [13].

1.3.2 Visible Light Communication

With the advent of the communication revolution the world has become truly a global village. However the only major threat to this steadfast revolution is the scarcity of spectrum in the RF range. The United Nations Population Fund forecasts that by 2030 approximately 60 per cent of the world population will live in an urban environment, while 27 megacities with greater than 10 million people are anticipated to exist [14]. Thus it has become a necessity rather than a luxury to find new spectrum in the electromagnetic region. As the infrared spectrum is already bursting at its seams, it is a natural choice to shift to the visible light region. The visible light spectrum has a few natural advantages:

1. This region has been left alone from regulation and has about 300 THz of untapped bandwidth
2. There is natural security as it is difficult to pick up a signal from a different room; the person has to be directly below the light source
3. There are no health regulations to restrict the transmitted power unlike the IR region
4. It is confined to a small geographical area
5. No interference from RF signals due to the large bandwidth present·

The recent advancement in LED technology has propelled a new generation of lighting and is replacing the older incandescent bulbs. The current luminous efficacy of LED lamps and luminaires is around 100 l m/W (lumens per Watt), and is expected to reach 200 l m/W around 2025, which is much higher than incandescent lamps (around 20 l m/W) and fluorescent lights (around 100 l m/W). LED lamps do not only have high luminous efficacy, but also long sustainability. LED lamps typically have a lifetime of 40,000 hours, which is 40 times longer than incandescent lamps. Due to all these reasons there is a renewed interest in VLC as these LED's can be instrumental in bringing about an all pervasive means of communication [15].

1.3.3 Intelligent Transportation System (ITS)

ITS have strong interfaces and overlapping with various other systems such as M2M, IoT, Smart Cities, Vehicular Communication, etc. The main aim of ITS is to keep the traffic and information both flowing simultaneously. It consists of smart infrastructure technologies implanted in traffic lights, vehicle parkings, toll booths, roads and bridges to facilitate them communicate among themselves and with the roadside infrastructure. It works in such way that the system should face less congestion, provide more safety and maintain continuous flow of useful and timely information. Instead of traffic signal controlling the vehicles, the vehicle itself can control the traffic signal in the ITS. In the developing countries, ITS can play very important role in conveying critical information to the surrounding vehicles and can contribute in saving lives of innocent citizens [16].

1.3.4 Introduction of Intelligence in Wireless Networks

Self-description of Smart objects with digital resources cause the limitation of input qualities due to their long established physical appearances. To rectify these drawbacks, the incapable representation of sound, video clips etc. are capable representations which make the smart objects accessible and interactive. In this paper two techniques are utilized, magic lens & Personal Projection. These techniques give the deeper insights on the positive and negative aspects of these techniques [17]. The ability to understand & recognize the important data and what is not important in the scene is an invaluable asset.

For robotic manipulation & navigation, discerning and localizing of queried objects in range is an important aspect. A new Point Pair Feature containing discriminative description inferred from a visibility context. The experimental results based on two methods, in terms of recognition and runtime performance [18, 19]. With the prevalence of cell phones and GPS technology the opportunities of using location based information have accelerated. However the mobile terminal restrict the use of one hand while user needs to keep a close watch on small display. To overcome this problem, the paper proposed a toe-input method which realizes haptic interaction, direct manipulation and floor projection in Wearable projection system [20].

1.4 Segmentation Strategy for FWE

For the success of FEW, segmentation is very much necessary. Roughly the FEW segmentation can consist of various subdivisions as shown in Figure 1.10. Air interface technology segmentation containing technologies like 2G, 3G, LTE-FDD, TD LTE, Mobile Core, 5G New Radio, etc. Access points, access point controllers, cloud radio network fronthaul come under submarket segmentation. Technology segmentation covers 3G packet core, HLR MSS, LTE EPC, WiMAX Mobile core, 5G next generation core, macrocell backhaul, etc.

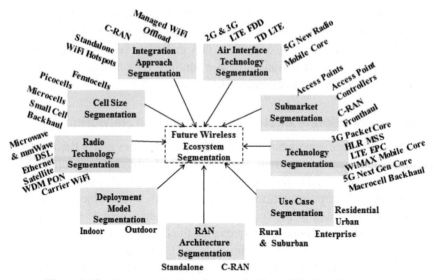

Figure 1.10   Segmentation Requirement for Future Wireless Ecosystem.

Managed WiFi offload, cloud radio access network, and standalone WiFi hotspots are the subcategories under integration approach segmentation. Cell size segmentation deals with Femtocells, Picocells, Microcells, and small cell backhaul. Microwave and mmWave, DSL, Ethernet, Satellite, WDM, PON and Carrier Wi-Fi fall under radio technology segmentation. Deployment model segmentation can have only two subdivisions like indoor and outdoor. Use case segmentation may have various kinds as rural, suburban, residential, urban, enterprise, etc. Segmentation is very important for the various subgroups to progress and grow faster [21].

1.5 Threats to Future Wireless Ecosystem

As the wireless networks are progressing with new technologies, so are growing the security and privacy threats in large number. Today's wireless world is a connected world. Everyone, everything, everywhere security threats are chasing the wireless networks in different ways as shown in Figure 1.11. Malwares are the Trojans and ransomware which can target the financial information and they make user to pay to unlock their system. Mobile threats consist of attacks on various applications installed on mobile, phishing attacks, and compromising the system access credentials like passwords, pins, and sensitive financial transaction information.

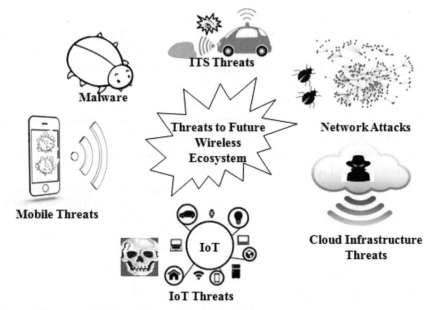

Figure 1.11 Future Wireless Ecosystem Threats.

IoT is the most attack targeted field by the cyber criminals. The sensors used to gather the information from their surrounding environment themselves are compromised and can be used for malign purpose. Cloud infrastructure is more prone to the distributed denial of service attacks. Network attacks are the most critical attacks implemented on the network protocols for execution of observation and intervention attacks [22].

1.6 Usable Security and Privacy

As wireless systems flourishing across a globe there are certainly many challenges coming ahead in future networks and technologies supporting innovative communication. In next generation machine to machine communication for sake of mankind will create fully interconnected network and develops distinct relationship between end users, consumers and service providers. It rapidly changes the scenario of collaboration and competition of multiple network entities within one system. Words like 'convergence', 'ubiquitous', 'cloud RAN', 'network virtualization', 'software defined networks', are becoming very common for describing future networks and applications.

Networks in future, will involve many nodes co-operatively working and communicating efficiently to provide services and routing the data. That means it may consists of cloud of RAN, MIMO systems or distributed antenna systems for newer applications like body PAN, vehicular environment, smart cities, smart grids etc. However conventional physical layer technology still treats every individual point to point link separately. Except the destined signal all other remaining signals from other nodes are treated as interference. So it is necessary to shift this paradigm to create network aware physical layer, which will exploit these interfering signals to much more efficient and secure network environment.

Network aware physical layer will need to apply more complex networks, involving multiple relays, sources, destinations and more than two hops. Network coding plays important role to make physical layer aware with network. Network coding used in such wireless environment should not only be limited to throughput improvement but also to address certain practical issues.

Interdisciplinary approach, by knowing the functioning of physical layer and network layer with each other can reduce substantially the problems associated with wireless communication. Combining physical layer with network layer along with data linking for media access can improve overall network capacity.

1.7 Summary

Future wireless ecosystem demands the robust and embedded security which can be trustworthy for the user. The next generation wireless systems are going to be user centric so the cybersecurity provision should be able to gain trust from the users. Also the ecosystem should be energy efficient, cost effective and environment friendly. The convergence of various existing and next generation wired and wireless technologies demands the strong and self-healing networks.

References

[1] Vandana Rohokale, Neeli Prasad, Ramjee Prasad, "Cooperative Wireless Communications and Physical Layer Security: State of the Art", Journal of Cyber Security and Mobility", Vol. 1, pp. 227–249, 2012.
[2] Alexander Graham Bell – First Telephone Patent #174465, 1876.

[3] Alexander Graham Bell, "Upon the production and reproduction of sound by light", Journal of the society of Telegraph Engineers, Volume-9, Issue-34, 1880, pp. 404–426.

[4] Gabriele Falciasecca, Barbara Valotti, "Guglielmo Marconi: the Pioneer of Wireless Communications", Proceedings of the 39th European Microwave Conference, 2009, pp. 544–546.

[5] Tom Farley and Mark Van Der Hoek, "Cellular Telephone Basics", Article posted on Privateline, Jan. 2006.

[6] Dholakiya J. H., Jain V. K., "Technologies for 3G Wireless Communications", International Conference on Information Technology, Coding and Computing, 2001, pp. 162–166.

[7] Wang, M. M., "Advanced Technologies in Evolved 3G Wireless Communication Systems", Keynote Talk at International Conference on Advanced Technologies for Communication, ATC 2008.

[8] Tinatin Mshvidobadze, "Evolution Mobile Wireless Communication and LTE Networks", 6th International Conference on Application of Information and Communication Technologies (AICT) 2012, pp. 1–7.

[9] Ramjee Prasad, Luis Munoz, "WLANs and WPANs towards 4G Wireless", The Artech House Universal Personal Communications, Boston, London, 2003.

[10] Savo Glisic, Beatrizn Lorenzo, "Advanced Wireless Networks: Cognitive, Cooperative and Opportunistic 4G Technology," Wiley Publications, Second Edition, June 2009.

[11] R. Prasad, P. Pruthi and K. Ramareddy, "The top 10 list for terabit speed wireless personal services", in Springer Journal on Wireless Personal Communications, Special issue on GIMCV, 2009.

[12] Soumya Kanti Datta, Amelie Gyrard, Christian Bonnet, Karime Boneoudaoud, "M2M Architecture Based User Centric IoT Application Development", 3rd International Conference on Future Internet of Things and Cloud (FiCloud), 2015.

[13] Blog on IDG Connect, "How IoT is vital for the ecological future of the connected planet", May 6, 2016. http://www.idgconnect.com/blog-abstract/16117/how-iot-vital-ecological-future-connected-planet

[14] Tanaka, Y., Komine, T., Haruyama, S., and Nakagawa, M. "Indoor visible communication utilizing plural white LEDs as lighting," in The 12th IEEE International Symposium on Personal, Indoor and Mobile Radio Communications (PIMRC 2001), San Diego, CA, Sept. 2001, pp. F81–F85.

[15] US Department of Energy, Energy Efficiency & Renewable Energy, Building Technologies Program, "Energy Savings Potential of Solid-State Lighting in General Illumination Applications", January 2012.

[16] John Walker, "Mobilizing Intelligent Transportation Systems (ITS) – GSMA Connected Living Programme", GSM Association Report, 2015.

[17] Fahim Kawsar, Enrico Rukzio and Gerd Kortuem "An Explorative Comparison of Magic Lens and Personal Projection for Interacting with Smart Objects" *MobileHCI'10,* September 7–10, 2010, Lisboa, Portugal.

[18] Alvaro Collet, Siddhartha S. Srinivasay, Martial Hebert "Structure Discovery in Multi-modal Data: A Region-based Approach" 2011 IEEE International Conference on Robotics and Automation Shanghai International Conference Center May 9–13, 2011, Shanghai, China.

[19] Eunyoung Kim and Gerard Medioni "3D Object Recognition in Range Images Using Visibility Context" 2011 IEEE/RSJ International Conference on Intelligent Robots and Systems, September 25–30, 2011. San Francisco, CA, USA.

[20] Daiki Matsuda, Keiji Uemura, Daiki Matsuda, Nobuchika Sakata, "Toe Input with Mobile Projector and Depth Camera" IEEE International Symposium on Mixed and Augmented Reality 2011 Science and Technology Proceedings 26–29 October, Basel, Switzerland.

[21] Telecommunications and Computing Blog, "The Wireless Network Infrastructure Ecosystem: 2017–2030 Macrocell RAN, Small Cell, C-RAN, RRH, DAS, Carrier Wi-Fi, Mobile Core, Backhaul and Fronthaul", Aug. 2017.

[22] CTIA Everything Wireless Report, "Protecting America's Wireless Networks", April 2017.

Biographies

Vandana Milind Rohokale received her B.E. degree in Electronics Engineering in 1997 from Pune University, Maharashtra, India. She received her Masters degree in Electronics in 2007 from Shivaji University, Kolhapur, Maharashtra, India. She has received her PhD degree in Wireless Communication in 2013 from CTIF, Aalborg University, Denmark. She is presently working as Professor in Sinhgad Institute of Technology and Science, Pune, Maharashtra, India. Her teaching experience is around 21 years. She has published one book of international publication. She has published around 30 papers in various international journals and conferences. Her research interests include Cooperative Wireless Communications, AdHoc and Cognitive Networks, Physical Layer Security, Digital Signal Processing, Information Theoretic security and its Applications, Cyber Security, etc.

Ramjee Prasad is a Professor of Future Technologies for Business Ecosystem Innovation (FT4BI) in the Department of Business Development and Technology, Aarhus University, Denmark. He is the Founder President of the CTIF Global Capsule (CGC). He is also the Founder Chairman of the Global ICT Standardisation Forum for India, established in 2009. GISFI has the purpose of increasing of the collaboration between European, Indian, Japanese, North-American and other worldwide standardization activities in the area of Information and Communication Technology (ICT) and related application areas.

He has been honored by the University of Rome "Tor Vergata", Italy as a Distinguished Professor of the Department of Clinical Sciences and Translational Medicine on March 15, 2016. He is Honorary Professor of University of Cape Town, South Africa, and University of KwaZulu-Natal, South Africa. He has received Ridderkorset af Dannebrogordenen (Knight of the Dannebrog) in 2010 from the Danish Queen for the internationalization of top-class telecommunication research and education.

He has received several international awards such as: IEEE Communications Society Wireless Communications Technical Committee Recognition Award in 2003 for making contribution in the field of "Personal, Wireless and Mobile Systems and Networks", Telenor's Research Award in 2005 for impressive merits, both academic and organizational within the field of wireless and personal communication, 2014 IEEE AESS Outstanding Organizational Leadership Award for: "Organizational Leadership in developing and globalizing the CTIF (Center for TeleInFrastruktur) Research Network", and so on.

He has been Project Coordinator of several EC projects namely, MAGNET, MAGNET Beyond, eWALL and so on. He has published more than 30 books, 1000 plus journal and conference publications, more than 15 patents, over 100 PhD Graduates and larger number of Masters (over 250). Several of his students are today worldwide telecommunication leaders themselves.

2

Wireless Access in Future Smart Cities and Data Driven Business Opportunities

Vladimir Poulkov

Faculty of Telecommunications, Technical University of Sofia,
8 Kl. Ohridski Blvd., 1756 Sofia, Bulgaria

Abstract

The future of a Smart City (SC) will be shaped by the digitization and connection of everyone and everything with the goal of automating our life, maximizing the efficiency of what we do, augmenting our intelligence with knowledge that expedites and optimizes decision-making and everyday routines and processes. This chapter brings up the vision that the Access Network (AN) in SC will develop towards a flexible and cognitive Unified Wireless Access (UWA) infrastructure with intelligence spread down to the level of each single user. The UWA with all its different types of users will function as a unique large scale complex system with integrated Communication, Navigation, Sensing and Services (CONASENSE). The characteristics and performance of such a UWA are further discussed and related to the ones of a Cyber-Physical System (CPS). The motivation to compare the performance of the UWA to a CPS is for the goal of analyzing possible business opportunities and models some of which today are discussed in the context of CPSs. CPSs with advanced CONASENSE and cloud technology can enable new information and data-driven business models and services, could transform businesses and create new business. Considering the UWA from the viewpoint of a large scale type of CPS it is envisaged in this chapter that the UWA can serve as a tool for enabling new services, business models and opportunities based on data. The monitoring and storing of "big data" as a result of the inherent CONASENSE infrastructure and operation of a UWA will bring new data driven business opportunities.

Towards Future Technologies for Business Ecosystem Innovation, 21–40.

Keywords: Wireless Access Networks, Smart Cities, Cyber Physical Systems, Data-Driven Business Models, Big Data, Data Analytics.

2.1 Introduction

The development of information and communication technologies leads to an environment *"in which we are always connected and can communicate with any system, device or person, allowing continuous and perpetual augmentation of our lives"* [1]. The future of a Smart City (SC) will be shaped by the digitization and connection of everyone and everything with the goal of automating our life, maximizing the efficiency of what we do, augmenting our intelligence with knowledge that expedites and optimizes decision-making and everyday routines and processes. In a future SC a user (human or machine) will be able to connect to the network seamlessly and utilize a practically infinite capacity provided through an effective utilization of the telecommunication access infrastructure. Hyper-connected telecommunication infrastructures and networks will be the *"single, most indispensable element of binding humans and individual societies, industries, economies, and humans, building an infrastructure of large-scale, complex and highly networked systems whose efficiency, sustainability and protection would require intelligent, interoperable and secure ICT solutions and novel business models"* [2]. In other words *hyper-connectivity* will be one of the major symptoms of access networks in future SC.

In this chapter the major characteristics of an access network (AN) in a future SC are considered. The AN in a future SC with all its different types of users will function as a unique large scale complex system with integrated Communication, Navigation, Sensing and Services (CONASENSE). It is argued that in a future SCs the ANs will transform towards a unified wireless architecture (UWA) with a cloud based virtual core and operating system and will acquire the major characteristics and some of the inherent features of a Cyber-Physical System (CPS). Considering that CPS are regarded as systems in which collaborating computational elements are controlling physical devices by exploiting the data gathered from different sensors and other devices, which operate in the environment of the devices [3], it could be concluded that such a UWA through all the users, devices and infrastructure will be able to collect and store an enormous amount of data.

Further it is envisaged that the capability of such an UWA in a SC to store and process data can serve as a tool for enabling new business models and for the realization of novel information and data-driven service opportunities.

In a SC scenario the UWA provider can become a data provider for different service organizations or to other customers. Due to the inherent business, services and operations, the UWA provider in a SC will collect and own a huge amount of data that can be sold to other stakeholders or be processed and used for providing new added-value services. The potentially valuable 'big data' compounded as a result of ensuring the wireless access in a SC will embed a big value potential that could be well commercialized. The availability of massive amounts of data collected from the different types of users, environment and the network operations, administration and maintenance (OAM) in the cloud of the UWA network will enable the creation of very innovative data driven services that could generate many new benefits. Even today data is considered as a resource that if correctly exploited and through the application of proper data driven business models (DDBM) can guarantee growth and profit. Collecting and processing data via the core business the access providers in the future SCs could find new business models and opportunities.

The chapter is organized as follows. In the next section are presented the essential characteristics and development of the wireless AN in a future SC. In section three the performance of the UWA is compared to the one of a CPS and the related data that is collected and stored in the process of their operation is considered. In section four the possibilities for enabling new business models based on data from the OAM of a UWA network are outlined. The last section concludes the chapter.

2.2 The Unified Wireless Access for Future Smart Cities

In a SC a highly developed and intelligent wireless access infrastructure will be needed for the connection all the different types of objects, sensors and user terminals. Nowadays the essential characteristics and development of the wireless AN in a SC are related to the integration of multiple wireless access technologies and implementation of sophisticated transmission techniques such as: massive MIMO, beamforming, coordinated transmissions, etc. To meet future traffic demands and achieve maximum network energy efficiency the use of the frequencies above 6 GHz is considered, leading to ultra-dense access networks based on small and heterogeneous cells [4]. In addition to this, intelligent resource and interference management approaches utilizing novel techniques such as antenna muting, sleep modes, energy optimized mixed line rates, dynamic routing, topology optimizations, etc., are in the stage of practical implementations. The implementation of ultra-dense access architectures will impose many issues and challenges that could also act as disruptive factors and/or driving forces for novel wireless AN approaches in a SC [5]. Some of these are related to the appearance of interference and power control problems, complicated handover and load balancing issues and the provision of the required network Quality of Service (QoS) [6, 7].

To solve these issues converged cell-less networks are considered as the replacement of the conventional celled architectures of the cellular networks. The idea is to support mobile users in SCs by the introduction of the principle of unassociated transmission between the Access Points (APs) and mobile user terminals, and introduction of flexible mobile user association schemes. In a conventional cellular network a user or mobile terminal is associated to only one AP (or base station) while in a cell-less network a flexible communication approach is realized with one or more APs. The latter is realized through the adaptive adjustment of the number of APs by the requirements of the mobile terminal, the respective applications and the wireless channel status in different environments. It is argued that such converged cell-less and cloud based software defined telecommunication networks (SDN) will provide flexible solutions in coverage and energy efficiency for future SCs. The cell-less telecommunication access approach can solve some of the challenges related to the physical level, but for the goal of performance optimization, user cooperation, mobility prediction and implementation of flexible solutions in coverage and energy efficiency, the intelligence has to be implemented overall in the cloud-based infrastructure of the network [8–10].

Bringing intelligence in HetNets is not a new topic and many Artificial Intelligence (AI) based techniques have been proposed for the implementation of autonomic capabilities for management, self-configuration, self-organization and self-optimization of HetNets. The tendency towards Self Organizing Networks (SON) is one step forward to a novel approach towards incorporating intelligence in the network, which will have influence also on the way communication networks are accessed. For sure, super-high speeds (or super-broadband) and service ubiquity will be some of the inherent features of the ANs of future SCs. For this to be ensured cognitive and autonomic networking features as the ones for SON have to be implemented, as for the realization of such ANs the intelligent provision of resources will be a must [11, 12].

The AN of future SCs should have all the characteristics of a network which is able to offer revolutionary services, capabilities, and facilities that are hard to provide using the existing network technologies.

Unlike the original IP based infrastructures focused on technical connectivity, routing, and naming, the scope of such ANs should encompass all levels of interfaces for services, manageability and technical resources (networking, computation, storage, control). Considering all mentioned above, it could be argued that in future SCs the ANs will develop beyond this practical cell-less approach, will transform towards a UWA telecommunication infrastructure and will acquire the characteristics and performance of CPSs. A justification of this assumption is made further below taking into consideration the perspectives for the development of future ANs outlined in [1, 13, 14] and the inherent features of CPS.

From technological, user and service provider perspective this should be a unified type of dynamic fixed and utilizing small cells and different types of wireless access technologies. Figure 2.1 is an illustrative example of the different types of wireless nodes, users and sensing equipment available as part of the infrastructure of the AN of a SC. The topology and user requirements awareness, the real-time sensing and data processing, and the possibility of self-organization and self-configuration, will ensure features such as: modularity and flexibility of the AN; possibility of efficient resource management; user discovery and matching; adaptation to the changes in users requirements; adaptation to changes in the networks operation status; etc. This all will allow implementation of flexible access policies for efficient and scalable utilization of all available communication resources. Mobility will be and inherent feature of the UWA related not

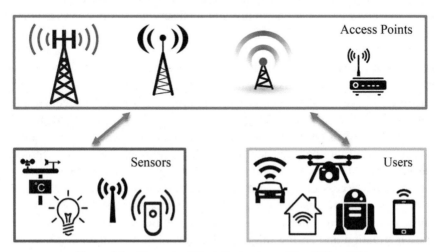

Figure 2.1 Elements of the infrastructure of the AN in a SC.

only to the support of mobile users and services in SC, but also movement of virtual resources such as computing resources, virtual access points and applications. Each virtual resource could be moved according to the users' demands and users could be dynamically attached or reattached to some virtual access point depending on the application characteristics. For this to be realized an Access Network Operating System (ANOS) will be needed. The ANOS will provide a centralized network abstraction that decouples the underlying specifics of network elements and domains from the network services [1].

The implementation of intelligence in the ANOS, such as the self-planning, self-organizing, self-monitoring, self-regulating properties, could be considered as addition of another technological dimension for ensuring the apparent infinite throughput, network capacity and service ubiquity. The ANOS will provide intelligence down to the level of each single user (human or machine) and enable the evolvement of more and more intelligent person-alized access solutions, taking into consideration user specific requirements and the personal characteristics of each user in the network. To fully imple-ment the intelligence in the system user behavior and user intent features will need to be captured, stored and analyzed. Big Data analytics and artificial intelligence (AI) approaches will be applied to forecast the characteristics of connectivity and traffic patterns and users' behavior in order to implement autonomic adaptation of the network performance to the rapidly changing customer demands and service attributes. This will allow every service to be personalized based on the understanding the specific user-service needs; to be accessible and easy to use and to provide such service experience as expected by the user.

While today's solutions in most cases interpret the term "user" as a human using mobile device, this will have to be extended to include non-human entities such as autonomous cars, unmanned aerial vehicles, robots, devices, machines etc., in which more and more intelligence is implemented. Besides that such users will be generating enormous number of entries in the AN that will exceed in orders the human users, also the characteristic features of their connectivity patterns and user behavior in many cases can approach and be similar to the ones of humans. The integration and virtualization of resources in "access clouds" and implementation of AI in a virtual core to process the massive amount of user generated data, on top of a sophisticated physical access will infer the best communication scenarios for each user in each situation and ensure ultra-low-latency, high reliability, and user and service scalability.

As reasoned in [14], there is a believe that a new, unified access communications paradigm will emerge in the future suggesting "*a convergent future for all types of communications, which in turn suggests that unifying communications platforms will emerge as a critical technology enabler that seamlessly and conveniently unifies all communication modes*". It should be considered that the UWA, which will be the most suitable for future SCs, has to be based on a "*unique virtualized access infrastructure and APs, based on novel and sophisticated and transmission methods, organized through clear rules for hierarchy, territory, resource allocation, spectrum utilization. This will be a shared-facilities based access network with shared virtual access infrastructures and shared virtual cores, based on common approaches for virtualization of technologies and services, fully IP based and with cross-layered access "virtualizing" all different networks and substructures (operators, private networks, etc.)*" [14].

The structure of a sample UWA network is shown in Figure 2.2. The users' (human and non-human) actions, related to their behavior (mobility, connectivity, requirements, generated traffic, etc.), are monitored via a User Action monitoring block collecting information from the user terminals and the UWA network. The resulting user action data is send to the Virtual Network Core (VNC). In the VNC the ANOS and other functionalities related to the intelligence of the network are implemented. To fully utilize the intelligence of the network the VNC receives data from monitoring of the overall operational state of the UWA network and other physical parameters (environmental, channel conditions, spectrum utilization, interference, etc.).

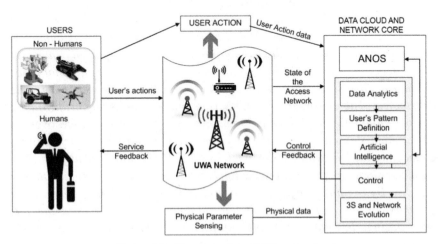

Figure 2.2 Structure of the UWA network.

Based on AI and 'big data' analytics actions are generated. Through respective feedback loops the operational state of the network is controlled and the required service provided to the users. The self-configuration, self-organization and self-optimization (3S) and network evolution features are implemented in the VNC. Due to virtualization and unification of the access, adaptation to an ever-changing physical environment and stochastic user behavior will be possible so that the system to be able to implement the concept of "hyper-connectivity".

2.3 Performance of the UWA as a Cyber Physical System and the Data in the Cloud

The structure of the UWA from Figure 2.2 implies that such an UWA in a SC with all its different types of users, actions, intelligence and feedback control will function as a unique large scale complex system based on CONASENSE. The characteristics and performance of such a system could be considered from the viewpoint of a CPS. The term 'Cyber-Physical System' describes hardware-software systems, which tightly couple the physical world and the virtual world. CPSs are software-intensive, intelligent systems with the capability to collaborate, adapt, and evolve. A typical CPS includes physical components, sensing network, computing devices, and communication network to acquire data streams from the physical world and enable the physical world to be monitored, controlled and influenced based on embedding intelligence, for the goal of adaptation, self-coordination and system evolution [15]. In [16] the following definition of CPSs is given: *"CPSs are physical, biological, and engineered systems whose operations are integrated, monitored, and/or controlled by a computational core. Components are networked at every scale. Computing is "deeply embedded" into every physical component, possibly even into materials. The computational core is an embedded system that usually demands real-time response, and is often distributed. The behavior of a CPS is a fully-integrated hybridization of computational (logical) and physical action."* A very extensive overview of different definitions of CPS is given in [17]. An up-to-date definition connecting the IoT with CPS is given in [18]: *"CPSs are systems featuring a tight combination of, and coordination between network systems and physical systems. By organic integration and in-depth collaboration of computation, communications and control (3C) technology, they can realize the real-time sensing, dynamic control and information services of large engineering systems".*

Some of the primary characteristics of CPS could be mapped to the elements and interactions of the UWA from Figure 2.2, such as: input and possible feedback from the physical environment; distributed management and control; real-time performance requirements; geographical distribution; multi-scale and systems of systems control characteristics; dynamic reconfiguration of the overall system on different time-scales; continuous evolution of the overall system during operation; possibility of emerging behaviors. As well as a CPS the UWA must also be able to tolerate failures, noise, uncertainty, imprecision, security attacks, lack of perfect synchronization, scale, openness, increasing complexity, heterogeneity and disconnectedness. Some of these potential problems are connected to non-technical savvy users in the loop of the CPS – humans. The humans will be either part of the loop (CPS with human in the loop [19]) or they can be in the center (human centric CPS [20]), but either way the human behavior imports uncertainty and unpredictability in the system [21]. The way to handle such a potential is by the incorporation of intelligence in the CPS in order to be able to execute complex tasks in dynamic environments, under unforeseen conditions and to allow adaptation and evolution. It is expected that CPSs will bring a revolution in monitoring, adaptively and intelligently controlling and influencing the physical world around us.

The motivation to compare the performance of the UWA to a CPS is for the goal of analyzing possible business opportunities and models some of which currently are considered in the context of CPSs. In an industrial context, the application of the CPSs concept is predominantly driven by initiatives like the "Industrial Internet" [22] and the Industry 4.0 initiative [23–26]. From an industrial viewpoint CPSs are considered to be some "smartified" equipment or industrial system [27] with capabilities of monitoring, control, optimization, and automation [28]. In [29] CPS are defined as systems which record physical data using sensors, affect physical processes through actuators, evaluate and save the recorded data, and via communication facilities (wireless and/or wired, local and/or global actively or reactively) interact both with the physical and cyber world. The Industry 4.0 concept is based on big and new types of data available in real-time, 'big data' that is becoming available from the operation of CPSs. CPSs allow the collection of huge amounts of data in real-time. This makes it possible to apply different data analytic approaches, for example for the goal of optimizing operations, introducing preventative maintenance or automating decision-making processes. In [28] and [30] the authors conclude that Industry 4.0 is about data and the 'big data' from CPSs will be a driver for new innovative business models and businesses which will create value in new innovative ways embracing the new possibilities emerging from the use of new technologies [31–34]. The results from the studies [31, 35, 36] show that the foundation of Industry 4.0 is data and that the new technologies made possible through CPS open up for new business models which will transform available data into value for the whole value chain. Creating value from data utilization is not a new business concept but CPSs with the ability to collect and analyze data in real time will open new possibilities for data collection and data utilization through their inherent features and system interactions.

The data collected in the process of operation of a CPS could be utilized in different ways. Embedding sensors in the components and augmenting CPSs to generate data is the most common pattern for utilizing data. Through embedded sensing, CPS enable extensive and intensive monitoring of their operation and environment, introduce remote controlling capabilities, enables optimization and the introduction of autonomous functions. The combination and correlation of data from different operations gives the possibility processes from the physical world to be modeled in virtual reality for the goal of implementing 3S features.

In [31] four types of data utilizations are discussed: monitoring, controlling, optimizing, and automating. The most fundamental usage of information is condition monitoring where preventative maintenance can be offered based on monitoring the condition of machines and products. Remote controlling is the second level of data utilization. If data collection and monitoring is combined with advanced data analytics, the third and fourth types of data utilization is reached where the 3S features could be implemented and fully autonomous systems realized. If the performance of a system could be monitored and has a feedback allowing remote control it can be optimized and made fully autonomous.

In [37] a five stage implementation architecture of CPS in an Industry 4.0 manufacturer is introduced. The first stage is called the connection level and is related to the process of data collection enabled by the connectivity of the machines and products. The second stage is related to converting big data to smart data, while the third one is called Cyber and represents a hub where all information is gathered. Stage four and five are the Cognition and Configuration stages, i.e. the data presentation and the feedback loop respectively. Considering Figure 2.2 it could be seen that the UWA has similar implementation architecture and data utilization features related to monitoring, controlling, optimizing, and automating as presented in [31].

In the UWA the monitoring capabilities of the architecture give the possibility to analyze the current status of all systems and equipment and are fundamental for realizing all kinds of user access scenarios. Monitoring enables the collection of different types of data through sensors and other external sources serving for the goal of being aware of the condition of the UWA and allows the access provider to ensure optimal functionality of the whole system and anticipate events or malfunctions. The controlling capabilities build upon the monitoring allow for keeping track not only of the current status of whole UWA but also interfere actively in OAM. The combination of monitoring the data generated by UWA and the control and service functionalities allow optimizing OAM. This presupposes a high degree of integration in the UWS in a similar way as in large scale CPS, as well as advanced analytical capabilities to derive actionable insights from the data collected from the operation of the system [38]. Due to the intelligence incorporated in the virtual core the system can learn about its operational environment, self-diagnose its own service and OAM needs, and adapt to the preferences of not only to the access provider but also to the preferences of the user for implementing different use scenarios and ensuring the desired QoS, thus having the possibility of fully autonomic functions and operation.

Considering the UWA in a SC from the viewpoint of a large scale CPS it will be interesting to consider the huge amount and different types of data that will be sensed, generated and stored in the cloud of the system. In general there will be two types of data. The first one is the internal or operational system data collected from all the equipment in operation in the access network. The second is the external data which is needed for optimizing performance of the UWA, and the implementation of intelligence and autonomic capabilities for management, self-configuration, self-organization and self-optimization (3S) of the network. If we reason in the direction what kind of external data will be needed to incorporate these capabilities of the network we could come to a general conclusion that almost any kind of data that could be collected in relation to the environment and the users in the network will be relevant. For example, environmental and wireless channel dynamics data, spectrum utilization data and user mobility and distribution data is needed for the implementation of the 3S features of the UWA network [11]. This will include historical data such as: user (human and non-human) service requirements, statistical data related to number of users connecting to an access node in time, user mobility data, spectrum occupancy data, hot spots, heat and interference maps, OAM related data, etc. Correlating and implementing of machine learning [39] and data analytics algorithms on all the data in the cloud of the UWA network could benefit the OAM and the efficient performance of the network. It will allow more complex services and applications to be enhanced and offered to the users. As mentioned above, the application of advanced data analytics on the data collected in the cloud the fourth type of data utilization could be implemented, i.e. realization of autonomic features in the UWA network.

The data analytics basically are three types: descriptive, predictive and prescriptive. Descriptive analytics use data aggregation and data mining to provide insight into the past and to better understand changes that have happened in the system. With techniques such as statistics, modelling, data mining, and machine learning predictive analytics is most suitable for making predictions on the system's OAM. Prescriptive analytics which use optimization and simulation algorithms to advice on possible outcomes is about being able to analyze past and thus forecast possible future states and operating scenarios and on that basis recommending actions for more effective operation or negative impact avoidance. Besides the assistance in the autonomous and 3S operation of the network the application such data analytics approaches could help to obtain insights about the performance of the overall access network, user behavior and requirements in order to make appropriate decisions and ultimately gain a competitive advantage on behalf of the access providers.

Example of application of data analytics on collected RF (Radio Frequency) data for the goal of interference management is the implementation of a RF data analytics approach on spectrum monitoring data as described in [40]. RF data analytics is the analysis to uncover what is buried in the data by mining massive datasets of RF data at different resolution levels. The descriptive analytics can be used for example for statistical analysis of user activity, channel occupancy, user distribution, interference management etc. Prescriptive RF data analytics allows optimization of spectrum usage and interference avoidance, while predictive analytics could be used to predict near-term or long term spectrum usage and/or variations in the power or frequency of wanted and unwanted signals.

Other examples that could be seen already today are in the surge of data and multimedia traffic which puts huge challenges for mobile operators to allocate resource effectively and to improve energy efficiency. Early detection and prediction of data traffic not only can help mobile operators anticipate upcoming congestions, but can also build a model for resource allocation to improve network capacity. As over 80% of the power consumption comes from their mobile nodes (base stations), traffic prediction also has economic and ecological benefits for mobile operators to reduce power consumption applying "sleeping modes" at their mobile nodes when the traffic could be neglected [41]. Predicting traffic density in different areas in working days and holidays is also applied nowadays [42] based on the analysis of the billing data or the "Call Detail Record" that the mobile operators collect. By examining this data, it is possible to analyze the activities of the people in urban areas and discover the human behavioral patterns of their daily life. These datasets can be used for many applications that vary from urban and transportation planning to predictive analytics of human behavior [42].

2.4 The UWA as Enabler of New Data Driven Business Models in SC

CPSs with advanced communications and cloud technology can enable new information- and data-driven services and transformation of existing business models. DDBM could transform businesses and create new business. Even today examples could be found where the data as well as insights obtained from the CPSs operations are used to realize different data-driven service opportunities. In such scenarios, traditional manufacturing organizations are becoming data providers for service businesses or different stakeholders [30]. Considering the UWA from the viewpoint of a CPS, besides the new technical capabilities and available data in the cloud of the UWA, such a large scale type of CPS also can serve as a tool for enabling new services, business models and opportunities. The monitoring and storing data as a result of the OAM of a UWA will generate 'big data' which can bring new innovative ways for creating value and new business models.

Since the appearance of 'big data', the concept of data as 'the new oil' [43] appeared, looking at data as a natural resource that has to be utilized and refined to ensure profit. 'Big data' is defined as *'high-volume, high-velocity and high-variety information assets that demand cost-effective, innovative forms of information processing for enhanced insight and decision making'* [44]. This definition considers data from three aspects – volume of data, variety of different data types, and data velocity, the latter referring to the speed at which the data is created, processed and analyzed. There is also another fourth aspect addressing the uncertainty of the data: veracity [45]. Veracity refers to the question of the accuracy and reliability of a certain data type. The UWA architecture for sure will generate huge volumes and varieties of operational, user and environmental 'big data' with high velocity and veracity, and in addition will ensure the high connectivity and security for its distribution. The available and potentially valuable 'big data' compounded from a UWA has an embedded value potential that can enable new DDBM. A provider of UWA in a SC has the possibility to collect and aggregate data from a vast number of different types of sources of data.

DDBMs rely on data as a key resource and its collection and aggregation [46]. But the big business enabler is in the analytics ecosystem where two basic types of DDBM are considered: Data-as-a-Service (DaaS) and Analytics-as-a-Service (AaaS). The DaaS business model is based on creating value by collecting and/or aggregating data from a vast number of different data sources and performing statistical analysis on the collected or aggregated data. A big opportunity for the implementation of DDBM using the data in the cloud of the UWA is the provision of AaaS. The application of advanced descriptive analytics, predictive or prescriptive analytics on the data in the cloud of the UWA, together with the inherent capability of providing secure distribution or access to the results, could create value and bring revenues.

Justification of the fact that the value of data is not in the mere presence of data but in the ability to exploit it is presented in [47], where is envisaged that the underlying capabilities for the creation of value from 'big data' are related to people, systems, processes and organization. The importance of data integration and the provision of an integrated data ecosystem allowing the input and storage of data, the analysis of data from multiple sources, real-time collection and accessibility of data, are fundamental for creating value from data and the realization of DDBM. Such an integrated data ecosystem for sure will be the virtual core and the cloud of the UWA Network in a smart city.

The advantage and challenges in the creation of novel DDBM based on big data collected in the cloud of the UWA lie in the ability to collect and analyze data in real time and the fact that the additional costs associated with collecting and storing data for the creation of data-driven businesses will be practically zero. The costs for communications, security, interfaces and the distribution of the data and data related information will be also negligible as all these functionalities will be inherent features implemented in the UWA architecture and all the data will be available in its cloud.

2.5 Conclusion

This chapter considered the development of the AN in a future SC towards a flexible, cognitive and intelligent unified wireless communication access infrastructure. The characteristics and performance of such a UWA with integrated CONASENSE were discussed and related to the ones of CPSs. Considering the UWA from the viewpoint of a CPS it is envisaged that the UWA in a similar way as a large scale type of CPS will enable new services and business opportunities based on data. The monitoring and storing of "big data" as a result of the inherent communication infrastructure and operation of a UWA will bring new data driven business opportunities. The UWA architecture for sure will generate huge volumes and varieties of operational, user and environmental 'big data' with high velocity and veracity, and in addition will ensure the high connectivity and security for its distribution. The available and potentially valuable 'big data' compounded from a UWA has an embedded value potential that can enable new DDBM. A provider of UWA in a SC has the possibility to collect and aggregate data from a vast number of different types of sources of data. Correlating and implementing of machine learning and data analytics algorithms on all the data in the cloud of the UWA network will allow the creation of a data analytics ecosystem where business models based on Data-as-a-Service (DaaS) and Analytics-as-a-Service (AaaS) will be enhanced. The application of advanced descriptive, predictive or prescriptive analytics on the 'big data' in the cloud of the UWA, together with the inherent capability of providing secure distribution or access to the results, could create new value and bring revenues. An integrated data based ecosystem such as the virtual core and the cloud of the UWA Network in a SC, allowing the input and storage of data, the analysis of data from multiple sources, real-time collection and accessibility of data, will be the basis for creating value from data and the realization of novel DDBM.

Acknowledgments

This work was supported in part by the contract DN 07/22 15.12.2016 of the Bulgarian Research Fund.

References

[1] Weldon, M. (2016). The Future X Network: A Bell Labs Perspective. Boca Raton: Taylor & Francis Group. ISBN: 978-1-4987-5927-4.
[2] Prasad, R. (2012). Future networks and technologies supporting innovative communications. In IEEE International Conference on Network Infrastructure and Digital Content (IC-NIDC), September 21–23 (pp. 4–6). doi: 10.1109/ICNIDC.2012.6418846.

[3] Lee, EA. (2008). Cyber Physical Systems: Design Challenges. In IEEE International Symposium on Object Oriented Real-Time Distributed Computing (ISORC), Orlando, USA, May 5–7 (pp. 363–369).

[4] Cimmino, A., et al. (2014). The role of Small Cell Technology in Future Smart City Applications. Transactions on Emerging Telecommunications Technologies, 25(1), 11–20.

[5] Kyoseva, T., Poulkov, V., Mihaylov, M., and Mihovska, A. (2014). Disruptive innovations as a driving force for the change of wireless telecommunication infrastructures. Wireless Personal Communications, 78(3), 1683–1697. doi: 10.1007/s11277-014-1902-0.

[6] Han, T., et al. (2015). Interference Minimization in 5G Heterogeneous Networks. Mobile Networks and Applications, 20(6), 756–762.

[7] Giust, F., et al. (2015). Distributed Mobility Management for Future 5G Networks: Overview and Analysis of Existing Approaches. IEEE Communications Magazine, 53(1), 142–149.

[8] Han, T., et al. (2017). 5G Converged Cell-Less Communications in Smart Cities. IEEE Communications Magazine, 55(3), 44–50.

[9] Han, T., et al. (2014). Mobile Converged Networks: Framework, Optimization, and Challenges. IEEE Wireless Communications, 21(6), 34–40.

[10] Xiaofei Wang, Xiuhua Li, and Victor C. M. Leung (2015). Artificial Intelligence-Based Techniques for Emerging Heterogeneous Network: State of the Arts, Opportunities, and Challenges. IEEE Access, 3, 1379–1391.

[11] Semov, P., Al-Shatri, H., Tonchev, K., Poulkov, V., and Klein, A. (2017). Implementation of Machine Learning for Autonomic Capabilities in Self-Organizing Heterogeneous Networks. Wireless Personal Communications, 92(1), 149–168. doi: 10.1007/s11277-016-3843-2.

[12] Peng, M., Liang, D., Wei, Y., Li, J., and Chen, H.-H. (2013). Self-configuration and self-optimization in LTE-advanced heterogeneous networks, IEEE Communications Magazine, 51(5), 36–45.

[13] Asenov, O., and Poulkov, V. (2014). Towards a Unified Virtual Mobile Wireless Architecture. Journal of Communication, Navigation, Sensing and Services (CONASENSE), 1(1), 93–104. ISSN: 2246–2120.

[14] Poulkov, V. (2016). Beyond the Next Generation Access. In R. Prasad, S. Dixit, (Eds.), Wireless World in 2050 and Beyond: A Window into the Future! (pp. 17–39). Switzerland: Springer. ISBN: 978-3-319-42140-7.

[15] Geisberger, E., and Broy, M. (Eds.). (2012). AgendaCPS: Integrierte Forschungsagenda Cyber-Physical Systems. Springer. doi: 10.1007/978-3-642-29099-2.

[16] Gill, H. (2008). A Continuing Vision: Cyber-Physical Systems. In Annual Carnegie Mellon Conference on the Electricity Industry Future Energy Systems: Efficiency, Security, Control, March 10–11.

[17] Engell, S. (2014). Cyber-physical Systems of Systems – Definition and core research and innovation areas. Working Paper of the Support Action CPSoS. http://www.cpsos.eu/wp-content/uploads/2015/07/CPSoS-Scope-paper-vOct-26-2014.pdf.

[18] Liu, Y., Peng, Y., Wang, B., Yao, S., and Liu, Z. (2017). Review on Cyber-physical Systems, IEEE/CAA Journal of Automatica Sinica, 4(1), 27–40.

[19] Feng, S., Quivira, F., and Schirner, G. (2016). Framework for Rapid Development of Embedded Human-in-the-Loop Cyber-Physical Systems. In IEEE International Conference on Bioinformatics and Bioengineering (BIBE), Oct 31–Nov 2 (pp. 208–215). doi: 10.1109/BIBE.2016.24.

[20] Romero, D., Bernus, P., Noran, O., Stahre, J., and Fast-Berglund, Å. (2016). The Operator 4.0: Human Cyber-Physical Systems & Adaptive Automation towards Human-Automation Symbiosis Work Systems. In International Conference Advances in Production Management Systems (APMS), Sept 3–7 (pp. 677–686).

[21] Manolova, A., Poulkov, V., and Tonchev, K. (2017). Challenges in the Design of Smart Vehicular Cyber Physical Systems with Human in the Loop. In L. P. Ligthart, R. Prasad (Eds.), Breakthroughs in Smart City Implementation (pp. 165–186), Aalborg, Denmark: River Publishers. ISBN: 978-87-99932-72-4.

[22] Evans, P., and Annunziata, M. (2012). Industrial Internet: Pushing the Boundaries of Minds and Machines. General Electric.

[23] Böhmann, T., Leimeister, JM., and Möslein, K. (2014). Service Systems Engineering: A Field for Future Information Systems Research. Business & Information Systems Engineering, 6(3), 73–79.

[24] Matzner, M., and Scholta, H. (2014). Process Mining Approaches to Detect Organizational Properties in Cyber-Physical Systems. In European Conference on Information Systems (ECIS), Tel Aviv, Israel.

[25] Soeldner, C., Roth, A., Danzinger, F., and Moeslein, K. (2013). Towards Open Innovation in Embedded Systems. In Americas Conference on Information Systems (AMCIS), Chicago, USA.

[26] Zdravković, M., Noran, O., and Trajanović, M. (2014). Towards Sensing Information Systems. In International Conference on Information Systems Development (ISD), Varazdin, Croatia.

[27] Mikusz, M. (2014). Towards an Understanding of Cyber-physical Systems as Industrial Software-Product-Service Systems. Procedia CIRP, 16, 385–389.

[28] Heppelmann, J.E., and Porter, M.E. (2014). How Smart, Connected Products Are Transforming Competition. Harvard Business Review, 92, 64–86.

[29] Acatech. (2011). Cyber-Physical Systems: Innovationsmotor für Mobilität, Gesundheit, Energie und Produktion. Heidelberg: Springer Verlag.

[30] Herterich, M.M., Uebernickel, F., and Brenner, W. (2015). The Impact of Cyber physical Systems on Industrial Services in Manufacturing. Procedia CIRP, 30, 323–328.

[31] Brettel, M., Friedrichsen, N., Keller, M., and Rosenberg, M. (2014). How virtualization, decentralization and network building change the manufacturing landscape: An Industry 4.0 perspective. Periodical, 8(1), 37–44.

[32] Kolberg, D., and Zühlke, D. (2015). Lean Automation enabled by Industry 4.0 Technologies. IFAC-PapersOnLine, 48(3), 1870–1875.

[33] Cao, G., Duan, Y., and Li, G. (2015). Linking Business Analytics to Decision Making Effectiveness: A Path Model Analysis. IEEE Transactions on Engineering Management, 62(3), 384–395.

[34] Geissbauer, R., Schrauf, S., Koch, V., and Kuge, S. (2014). Industry 4.0 – Opportunities and Challenges of the Industrial Internet. https://www.pwc.nl/en/assets/documents/pwc-industrie-4-0.pdf.

[35] Mittermair, M. (2015). Industry 4.0 Initiatives. SMT: Surface Mount Technology, 30(3), 58–63.

[36] Nathan, S. (2015). Getting to grips with Industry 4.0. Engineer (00137758), 296(7867), 30–34.

[37] Lee, J., Bagheri, B., and Kao, H. (2015). Research Letters: A Cyber-Physical Systems architecture for Industry 4.0-based manufacturing systems. Manufacturing Letters, 3, 18–23.

[38] Chen, H., Chiang, R.H., and Storey, V.C. (2012). Business Intelligence and Analytics: From Big Data to Big Impact. MIS Quarterly, 36(4), 1165–1188.

[39] Alpaydin, E. (2014). Introduction to machine learning. MIT press.

[40] Iliev, I., Bonev, B., Angelov, K., Petkov, P., and Poulkov, V. (2016). Interference Identification based on Long Term Spectrum Monitoring and Cluster Analysis. In IEEE International Conference "BlackSeaCom", Varna, Bulgaria, June 6–9 (pp. 1–6).

[41] Ni, F., Zang, Y., and Feng, Z. (2015). A study on cellular wireless traffic modeling and prediction using Elman Neural Networks. In International Conference on Computer Science and Network Technology (ICCSNT), December 19–20 (pp. 490–494).

[42] Khan, F. H., Ali, M. E., and Dev, H. (2015). A hierarchical approach for identifying user activity patterns from mobile phone call detail records. In International Conference on Networking Systems and Security (NSysS), January 5–7 (pp. 1–6).

[43] Rotella, P. (2012). Is Data The New Oil?. Forbes. http://www.forbes.com/sites/perryrotella/2012/04/02/is-data-the-new-oil.

[44] IT Glossary. (2012). Big Data. http://www.gartner.com/itglossary/big-data/.

[45] Schroeck, M., Shockley, R., Smart, J., Romero-Morales, D., and Tufano, P. (2012). Analytics: The real-world use of big data. How innovative enterprises extract value from uncertain data. IBM Institute for Business Value, Saïd Business School at the University of Oxford.

[46] Hartmann, P. M., Mohamed, Z., Feldmann, N., and Neely, A. (2016). Capturing value from big data – a taxonomy of data-driven business models used by start-up firms. International Journal of Operations & Production Management, 36(10), 1382-1406. doi: 10.1108/IJOPM-02-2014-0098.

[47] Verhoef, P., Kooge, E., and Walk, N. (2015). Creating value with big data analytics: making smarter marketing decisions. Routledge, Taylor & Francis Group. ISBN: 978-1-315-73475-0.

Biography

Professor Vladimir Poulkov PhD, received his MSc and PhD degrees at the Technical University of Sofia (TUS), Bulgaria. He has more than 30 years of teaching, research and industrial experience in the field of telecommunications, starting from 1981 as an R&D engineer in industry, and developing his carrier to become a full professor at the Faculty of Telecommunications, Technical University of Sofia. He has successfully managed and realised numerous industrial and engineering projects related to the development of the transmission and access network infrastructures in Bulgaria, as well as many R&D and educational projects. His fields of scientific interest and expertise are related to interference and resource management in NGN and IoT. He is author of more than 100 scientific publications and tutors BSc, MSc and PhD courses in Information Transmission Theory and Next Generation Access Networks. He is head of the "Teleinfrastructure R&D" laboratory at the Technical University of Sofia, Chairman of the Bulgarian Cluster of Telecommunications and Vice-Chairman of the European Telecommunications Standardization Institute (ETSI) General Assembly. He is a Senior IEEE Member.

3

5G and Telemedicine: A Business Ecosystem Relationship within CONASENSE Paradigm

Ambuj Kumar, Sadia Anwar and Ramjee Prasad

CGC, Department of Business Development and Technology,
Aarhus University, Herning, Denmark

Abstract

The use of smartphones has been increasing rapidly and it is expected that in future most people will have a smartphone capable of high speed Internet connection. The capability of smartphones with high definition display, computation power and multitude of sensors made it an excellent candidate for telemedicine application. Telemedicine's applications and high data medical information generally require high definition visuals and lower latency connection, in addition mobility and reliability. The next generation of wireless communication standard, known as 5G, will provide data speed in (Gigabit per second) Gb/s with lower latency and higher reliability connection, and can be better approach for future telemedicine. In this chapter we survey the current state of telemedicine along with examining the characteristics of 5G technology. We shall see how the Telemedicine and 5G, together, form a composite business ecosystem. We also discuss the research challenges concerning 5G and telemedicine.

Keywords: 5G, Telemedicines, Wireless Communications, Business Modeling, Business Ecosystems.

Towards Future Technologies for Business Ecosystem Innovation, 41–60.

3.1 Introduction

According to recent survey [1] 90% of medical practitioners are utilizing smartphones for clinical applications. By using smartphones, communication or interaction to patients becomes very easy and more effective. The estimated number of clinical applications available on smartphone is approximately 95,000. Furthermore, Internet of things (IoTs) is changing the way of our everyday life, as it is going to allow many things feasible and accessible e.g., smarter health care services, smart town and cities, smart automobiles and smarter retailing and shopping.

Wireless mobile communication is not only changing the way of our life but also becoming an integral part of our lives. Smartphones contribute a lot in keeping us updated in every aspect of life, as they can access a large number of information through Internet anytime, anywhere, and at a high speed. In early days, cellular technology was only focused on messaging and making phone calls but now it is possible to perform much more complex communication and computation tasks, in particular being connected with broadband mobile Internet connection.

As smartphones or smart devices are becoming more and more computationally powerful, along with high resolution displays, multiple ways of interaction, and various kind of embedded sensors that need for faster data rate, the low latency and higher reliability become more and more crucial and thus important. Therefore, 5G will be very important in this scenario, having capacity and high data rates, and is able to transmit data at Gb/s speed in 2020, which is about 200 times faster than 4G.

Telemedicine or telehealth is an inter-disciplinary area, which uses telecommunication technology to deliver medical information, medical help and services to remote areas at distant places, especially for the elderly and disabled persons. Telemedicine can be utilized as a first line strategy in case of emergency situations and disaster management. It was first applied in Mexico disaster by NASA in 1985. Voice communication was possible within 24 hours. With the passage of time and advancement in technology, this latency is now become small as in milliseconds by continuous modification and now by the new technology concept 5th generation mobile wireless communication (5G) which is becoming reality in 2020 [2].

We predict 5G along with telemedicine is going to revolutionize healthcare, and can be imagined in case of availability and time factors. Whereas, availability factor is basically introducing more specialists and medical physician to come and practice the telemedicine and time factor is handled by the faster wireless technology with high data rates such as 5G effectively and precisely for the delivery of virtual authentic medical care to patients.

The main lineaments of telemedicine are that it should be climbable, transparent or crystalline, provide geographical or global coverage, and fault tolerance with security, and authentication. Hospital digital networking concept came into being because of new faster wireless communication technologies, which are enhancing the approach of specialists and patients with each other to communicate visually and talk and share their condition with suggestive approach and strategies.

Business ecosystem [3] is a stage where various inter-/cross- disciplinary areas collocate for realistic commercial arena. Based on the aforementioned information, we can see that 5G and Telemedicine shall converge forming a steady and sturdy business ecosystems.

The rest of the chapter is structured as follows: Section 3.2 describes important attributes of 5G in relation to telemedicine. Section 3.3 presents what are the research challenges in connection to 5G and telemedicine. Section 3.4 shall elaborately stress on the Business domain of this amalgamation And lastly, conclude the chapter in Section 3.5.

3.2 Important Attributes of 5G In Relation to Telemedicine

In 2020, new and more revolutionized telehealth services will need wireless technology which can support high definition (HD) video quality, faster speed, low latency with no interruption in signaling pathway, more authentic in providing security and supportive for subscribers to use and subscribe applications especially in telemedicine, which is a concept truly based on the delivery of medical information to remote areas by utilizing wireless communication technology. 5G is going to address M2M (machine to machine) communication. Which is further divided into:

- Massive machine type communication, it includes low cost and low energy devices e.g. sensors involved in monitoring or diagnosis of vital signs. These devices need faster speed and real time communication [4].
- Mission critical type communication, it is real time controlled and has automated field function and processing such as robotically performed remote tele-surgery. In tele-surgery doctors can perform a surgery without their presence in a particular location, virtually through wireless technology. Tele-surgery needs more immense coordination between different surgeons which are remotely connected and this also requires

that information of each performed task should be available with minimum latency in real-time scenario. 5G, in this scenario, will be more reliable with faster speed and low latency [5].

Some of the important attributes concerning 5G are as follows:

3.2.1 Some of the Important Attributes Concerning 5G Are as Follows

- Connectivity and coverage in Gb/s [9], will be available even in natural disasters or emergency situations, thus having this technology, more lives can be saved.
- Strong authentication with secured patient profile would be imagined with 5G. However, research is still under process to avoid jamming attacks and similar complications.
- Remote monitoring of patients in distant areas would be easy, because 5G will be able to provide higher bandwidth.
- Power consumption of batteries is viewed for 2020, and charging is available for at least 10 years in M2M scenario, [9].
- 5G also enables various sensors and implant to connect with different service configurations to see patient condition and to provide them best regime or therapy.
- 5G has mobility with faster speed and low latency [9], which is less than one millisecond. Virtual reality and HD videos should be as detailed as human retina can detect. For this 300 Mb/s is generally required and this technology is 100 times above with low data rate and ultra-reliability, to carry out the High definition videos. It will be very helpful in dealing patients in video conferences to avoid breach in communication [6].
- Medical image transmission and videos require larger bandwidth. These images and videos are compressed during transmission and it is expected on the other end that there would be no data or image loss after receiving. 5G can handle HD videos and high data transmission in case of remote monitoring of patients, video conferencing, robotic equipment's, smart pharmaceuticals such as ventilators, fluid rate and drug delivery systems attached to the patients in remote areas. As an example, in diabetic patients, insulin reservoir are controlled and connected with wireless technology. Sensors provide the signals on daily basis for body glucose level. Insulin is then injected accordingly after taking the values.
- With 5G, first-aid could be delivered in small duration of time and communication without interruption available with the blink of eye.

- More and more patients are motivated to consult with doctors online. It is not only saving time but its cost effective features making it more reliable. 5G can play a more vital role to access these consultations frequently and remotely with ultra-low cost and low-end data rate reliability.
- Cloud service robotics for supportive therapy will be easy as offloading and uploading of information for recognition of language, object and cognitive skills will be predicted on real time with no interruption. Cost will also be reduced as less number of robots is involved [5].
- Remote monitoring of elderly patients will be optimized through continuous updates about life style modification and treatment for patients with chronic diseases and related disabilities. Their rehabilitation care plan and strategies can be reviewed by wireless communication, for this wireless technology has to follow the same standard for urban and rural areas where there is higher density of followers and users.

3.3 Research Challenges

A new and advance form of existing technology always has inherent challenges to deal with new problems. First generation mobile communication (1G) came in 1980s while fifth generation (5G) is going to be commercialized in 2020. It is a journey of approximately 40 years. During the modification and transformation of these technologies, we are still lacking proper infrastructure, standards, security and legislations concerning wireless communication technology.

Figure 3.1 shows and elaborate concept of telemedicine services running on 5G network. The challenges that we are facing in wireless communication concerning telemedicine are:

3.3.1 Bandwidth Requirements for Offloading Medical Services and Applications

Clinical services can be divided for telemedicine into two main categories within a clinical organization.

- Inter telemedicine services
- Intra telemedicine services

Tertiary clinical center implicate two or three medical professionals to deal with a common patient case especially in pediatric cardiology, where data

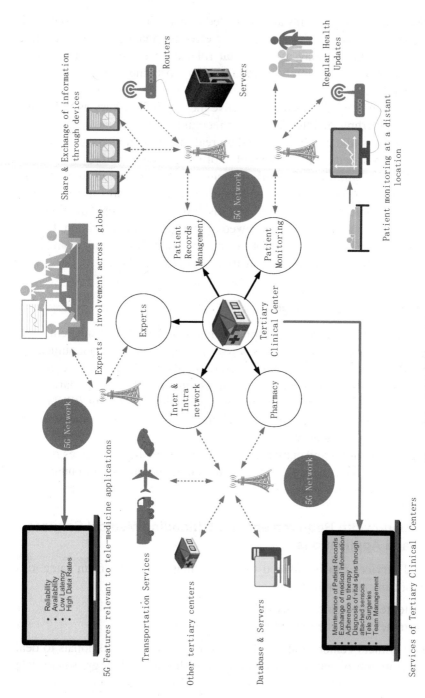

Figure 3.1   5G for telemedicine applications.

which is in the form of images or medical records are transferred to get a common solution of a given problem e.g., ECG, chest radiology and murmurs of neonates is called inter telemedicine services.

On the other hand, different tele conferences and meetings would be arranged to join different medical professionals from different departments called intra telemedicine services. This is mostly happened by communication through wireless technology. For Medical images transmission and videos require larger bandwidth and high resolution for 3D (three dimensional) images and other body scans so that after compression during transmission, there would be no data loss or interruption in the other receiving end because service is shared by different users at one time in a common place.

On the contrary, this band-width requirement provoke the situation to more critical level, when there is a need to see or monitor a patient on a distant place like patient with disabilities or elderly people with degenerative pathologies (e.g.: Alzheimer, or the insurgence of other pathologies such as stroke, cardio circulatory and muscular dysfunction etc. Last clinical advice is to perform physical activity regularly during the day and to maintain a social activity. For this there is a need of comprehensive health care plan. So, constant engagement with physician is needed.

The biggest socioeconomic challenge that has been arising in Europe is aging. According to EU commission public health policy, by 2025 more than 20% of Europeans will be 65 or over, with a particularly rapid increase in numbers of over-80s. (EU Commission) [7].

There is a need to decrease their abundance in hospital because it will not only create burden for medical staff but also made the environment more congested to deal with. Latest GPRS (General packet data rate service) system is not able to provide this real time communication precisely especially for the recording of patient movements during exercise, for this patient has to be monitored in hospital to get the values accurately. Apparatus utilized in hospitals are very outdated and based on the radiometry with limited area for coverage [8].

There is a need of high definition breakage free wireless communication that can fill up the gap of space (which means less occupancy of patients) and time factor (response is quick). Patient and physician can communicate data from different attached sensors, exoskeletons to see patient movements and smart pharmaceutical devices which are providing dose tailored on suggestive and maintained therapy on a signaling pathway by the physician on real time with no delay especially for remote areas. 5G can change this whole scenario and can help to fill this gap created by space and time factor. Data could be easily analyzed with suggestive therapy on a mobile anywhere, anytime with a few milliseconds delay and can be handled with fastest communication.

Less costly cellular displays, medical applications and software's can easily be accessed and managed. Medication and dose adjustment by sensors and smart pharmaceutical devices for cardiac and diabetic patients can easily be done by physician with positive feedback mechanism.

3.3.2 Reliability and Availability Requirements

Recent increase in emergency situation and disasters, forecasting or their prediction is not easy. Sometimes, it is very time consuming, but recently happened events, can give us a knowledge that how we can deal with them in future and how help can be provided based on a real time with no communication loss. Billons of people are affected by these disasters. These disasters could be technical, natural or human generated which can cause deaths, disabilities and psychological stress for a long time.

Hospitals are continuously active in these crises and the biggest challenge in disaster management is communication. Proper system and methodology is not fully available in handling, processing and transmission of critical data on real time especially at a distant place properly. In disasters, regional communication services are severely affected and communication becomes constrained and distorted.

There is a need of secure architecture for wireless communication system and 5G will be beneficiary in such scenario as 5G has a high band-width and low latency, and this system can be integrated with cloud computing system by this health care services will be more reliable and available.

3.3.3 Standards and Security

5G, being a new emerging technology is still lacking standards and this might take many months to years, to be defined as well implemented. Many of the universities and government organizations are working in this part to make 5G concept into a reality based technology. It is said that set standards will released in 2020, to make this target date into reality many of the public and government organization like IEEE, 3GPP, universities, ITU are making efforts to make this possible [8].

Prevailing Technological and operational guidelines which are in practice concerning telemedicine are [10]:

- Audio-video and data transmission should be of high quality and fulfill telehealth existing practice. Equipment's must have latest security software's as recommended by the device manufacturer. There must be a backup plan in case of communication or data loss.
- Multiple authentications and no activity timeout function can be utilized and in case of mobile or device loss or get stole, provider must be able to disable the connection and data.
- Real-time connectivity or synchronization requires bandwidth of at least 384kbps for down and uplink assignments and such kind of services must also provide a resolution of 640*480 at 30 frame/seconds. According to some health practitioner that high speed standard quality of video is not similar on a same bandwidth provided by different technologies.
- Video conference must utilize a link that in case of band width loss, stabilize and connection must not be lost.
- Whole disk encryption must be done for storage of a synchronized intentional data if cloud services are unable to give higher level of safety for health professionals and patient records, then it should be streamed to avoid unauthorized users and hacking attacks.
- Patient and physician should pretest the connection before communication is going to be happen. Tele or video conference software's must open one session to be conducted for discussion if there is an attempt or a hack to open second session, it must be automatically log off or access to second session must be denied.

These guidelines are helpful to set the standards. Here are some of the areas, where standardization is still needed.

3.3.3.1 Minimum delay require for delay application services

Telephonic education about medical issues require rate of data transfer from few 100 Mb/s to many 100 Gb/s The speed of 4G is approximately 10 to 20 Mb/s. Synchronized applications needs a minimum delay of 400 milliseconds [11].

International telecommunication service has given us the value of transmission of a data packet in IP networks for 3G which is 100 to 400 milliseconds for end to end delay and 1 per 1000 packets loss for synchronized services. In telemedicine for routine patient checkup QoS (quality of service) for 3G offers 150 to 400 millisecond delay which is less important as compared to emergency services where it offers a delay of 0 to 150 millisecond [12].

3.3.3.2 Bandwidth requirement for video based application

5G in such perspective is effective because it can transfer the images and data for telemedicine without interruption with very small delay or latency of about one millisecond. It will be more effective and reliable especially in remote patient monitoring with effective real time consultation [12].

Information technology utilization especially in health care sector has been very slow and the gatekeepers involved in this sector, making it difficult to develop in case of standards for data delivery and content [5].

In case of any network like Wi-Fi, 3G, 4G wireless network or wire system for an effective video call transmission, require a minimum band width of 230kbps (standard resolution) for uploading while for downloading of an additional video of ongoing call per video window require 178MB more. If we add more people in ongoing communication 128kbps are more needed for additional downloading per video window [14].

File size and type for transfer in medical facility is different for teledi-abetic retinopathy, MRIscans, digitalchest films, electrocardiogram studies and telepathology [15].

By adopting 5G which has data rate of 10 Gbit/s is more appropriate in transmission of these files with ease, higher speed and data rate because it has ability to cover all the challenges arises due to speed and audio video quality by providing higher bandwidth with low latency rates.

3.3.3.3 Security requirement and data encryption

Standards for data transmission and policies are not fully grown, still on basic or initial phase which needs reevaluation. There are not safety standards available for new innovative telemedicine services and for these cost is very high for their transmission on a broad band services [16].

Security makes a linkage between three main parts. These are self-sustainability which means, data is fully available not fragmented. Second is confidentiality which denotes that data is shared between authorized persons or users and third part of this linkage is availability which is the utilization of different services of networking. These three parts are interconnected and their balance actually is the success of provision of medical services. The first two parts are very important in telemedicine while, third is not as critical as others because there is always a backup plan available for data access. Self-sustainability is also important in this scenario, because in case of data loss it will directly affect the patient treatment or regime; the reason for this is that sound information is lost during transmission.

Data can also be encrypted, as a secret key is shared between one particular health unit and related staff dealing with a particular patient. They can access the data for treatment anytime. Data can be hacked or misused iso security is much more important within the hospital. Food and drug authority is only concerned to regulate medical instrument or devices and does not set any standards for consumer devices and application related to privacy concerns. So basically it is for patient safety but not concerned with patient privacy.

3.4 Business Ecosystem Paradigm in Relation to the Integration of Communication, Navigation, Sensing and Services

Telemedicine is a kaleidoscope of multi-/inter-/cross- disciplinary. The world telemedicine itself is intersection of two extremely independent scientific areas, namely telecommunications and medicine. This is unique to CONASENSE [17–19] as most of its works are associated to telecommunication paradigm. Through this discussion, we are pointing out the fact that CONASENSE is also associated to business ecosystem and telemedicine is one such example. Figure 3.2 shows the CONASENSE framework in

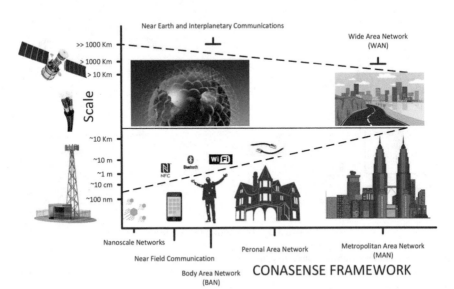

Figure 3.2 Conasense framework#.

Sources: #Nanobots: https://www.researchgate.net/topic/Network-Architecture;
Other pictures: https://stock.adobe.com

telecommunications. Here, we can see how communication technology ranges from Nanoscale Networks to Near Earth intra-/intra planetary communication, and no other field other than telemedicine has possibility of penetrating into the CONASENSE domain at every level of this framework. This means that from nanoscale to satellite communications, telemedicine can be benefitted at every level.

As an example, the nanoscale communications can be very beneficial in injectable nanobots who can travel in the blood vessels for pin-point treatments/surgeries, whereas the satellite scale can be used for broader medical services. Figure 3.3 shows that broader application of CONASENSE in relation to telemedicine, where, the human biological parameters are measure by biosensors that are either wearable or positioned across smart home. These biosensors use telecommunication to store and transmit parametric data. These transferred data can be then collected at the servers/computers for analysis, and/or can be used for calling appropriate services.

Figure 3.4 shows the hepta dimensional [20] business model that is uses the B-Lab, which is developed in CTIF Global Capsule, Aarhus University, Herning, Denmark. Figure 3.3 expresses telemedicine in CONASENSE perspective. Following is the discussion about the seven dimensions of telemedicine in relation to the CONSASENSE.

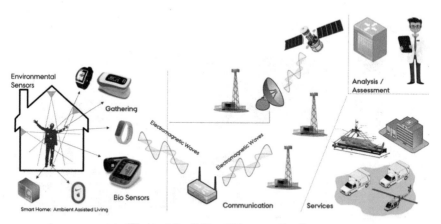

Figure 3.3   Telemedicine paradigm*.

Sources: *Blood pressure meter:
https://www.omron-healthcare.com/en/products/bloodpressuremonitoring;
Step sensor: Apple Nike and iPod Sensor; pulse oximeter: https://innovo-medical.com/products/ip900ap;
Generic pictures: Google search.

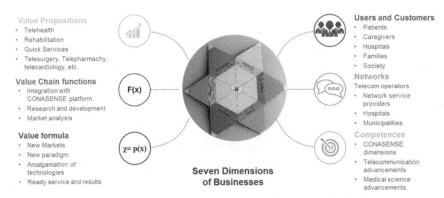

Figure 3.4 Telemedicine business model in seven dimensions, business ecosystem.

3.4.1 Value Proposition

It is the value that a business model provides to its customers or users. These values can be a product, service or solution of a problem. In case of telemedicine, the value propositions are Telehealth, Rehabilitation, Telepharmacy, Telesurgery, Telecardiology etc. This means that the telemedicine business models is incepted to intervene telecommunication between the patient and medical solution. Using the telecommunications, a patient can be facilitated with most of the medical needs from remote location including, rehabilitation, surgery, monitoring etc. Telemedicine can be extremely useful in underserved areas and in case of calamities.

3.4.2 Users and Customers

By definition, customers pay for the values and users return another value as compensation. In case of telemedicine, the customers can be hospitals, caregivers, municipality etc, whereas users may be doctors and end patients who do not pay directly to the telemedicine services. In both cases, the user/customers use telecommunications services to benefit their neets and the corresponsing provider uses them for prompt solutions.

3.4.3 Networks

This component correcponds to the relations with other businesses to achive values. In telemedicine, the Telecom providers, hospitals, caregivers are the networks that can be associated in the telemedicine process.

3.4.4 Competences

This dimension corresponds to the business competences. In case of telemedicine, all dimesions of CONASENSE shall be the begggest competences besides the medical advancements.

3.4.5 Value Chain Function

This dimension associates with other values that are inevitable to achieve the business values. Integration with CONASENSE platform is the most favourable function that can be assissted with market analysis and research and developments.

3.4.6 Value Formula

The value formula, meaning the returns of a business, of the telemedicine is the quality enhancements in patients' services. From rehabilitation to surgery, all the possible medical domains are redealy achievable due to the accomodation of CONASENSE paradigm.

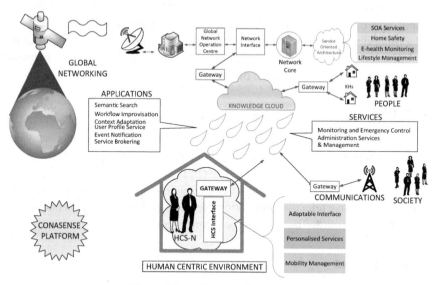

Figure 3.5   CONASENSE Platform.

3.4.7 Relations

This dimension relates all other dimensions with each other for the seamles business development.

Thus, we see that the Telemedicine-CONASENSE plaform, as shown in Figure 3.5 can be a boon to the society. By introducing human-centric features in the CONASENSE paradigm, a lot can be achieved in business ecosysmes.

3.5 Conclusion

In our survey we can conclude that work is still needed to define standards, and in particular security framework to make telemedicine services secure and also to make these services more applicable globally. A comprehensive framework is required to address future telemedicine using 5G technologies. There are, however, many uncertain areas what 5G is promising such as if a consistent, low latency and reliable connection will be available globally. More future research is needed to explore what kind of telemedicine application can emerge in connection to 5G technology, and more advanced smart devices having various sensors, and artificial intelligence. Further, it is discussed that how CONASENSE can be a key-player in application of telemedicine services by allowing all its verticals to percolate in telemedicine domain. We have also discussed the seven dimensions of telemedicine business ecosystem and how CONASENSE can play its role in the business enhancements.

References

[1] Mobile devices in healthcare come with pros and cons, SearchHealthIT. [Online]. Available: http://searchhealthit.techtarget.com/tip/Mobile-devi ces-in-healthcare-come-with-pros-and-cons. [Accessed: 03-Jun-2016].

[2] Smith, Y. A Brief History of NASAs Contributions to Telemedicine, NASA, 16-Aug-2013. [Online]. Available: http://www.nasa.gov/content/ a-brief-history-of-nasa-s-contributions-to- telemedicine. [Accessed: 03-Jun-2016].

[3] 5G Radio Access for Ultra-Reliable and Low-Latency Communications, Ericsson Research Blog, 11-May-2015.

[4] Lindgren, Peter. "Business model innovation leadership: how do SME's strategically lead business model innovation." International Journal of business and management 7.14(2012):53.

[5] 5G-PPP: 5G and e-Health.

[6] 5G Vision: 100 Billion Connections, 1 ms Latency, and 10 Gbps Throughput. [Online]. Available: http://www.huawei.com/minisite/5g/en/defining-5g.html. [Accessed: 03-Jun-2016].

[7] Policy – European Commission. [Online]. Available: http://ec.europa.eu/health/ageing/policy/index_en.htm. [Accessed: 03-Jun-2016].

[8] Oleshchuk, V., and Fensli, R. Remote Patient Monitoring Within a Future 5G Infrastructure, Wirel. Pers. Commun., vol. 57, no. 3, pp. 431439, Jul. 2010.

[9] Best, J. The race to 5G: Inside the fight for the future of mobile as we know it, TechRepublic. [Online]. Available: http://www.techrepublic.com/article/does-the-world-really-need-5g/. [Accessed: 03-Jun-2016].

[10] Bernard, J., and Linkous, J. D. Core Operational Guidelines for Telehealth Services Involving Provider-Patient Interactions. [Online]. Available: http://www.americantelemed.org/docs/default-source/standards/core-operational-guidelines-for-telehealth-services.pdfsfvrsn=6. [Accessed: 03-Jun-2016].

[11] Kota, S. L., Pahlavan, K., and Leppnen, P. A. Broadband Satellite Communications for Internet Access. Springer Science & Business Media, 2011.

[12] Y. 1541: Network Network performance objectives for IP-based services. [Online]. Available: https://www.itu.int/rec/T-REC-Y.1541-201112- I/en. [Accessed: 03-Jun-2016].

[13] 5G Radio Access for Ultra-Reliable and Low-Latency Communications, Ericsson Research Blog, 11-May-2015.

[14] What is the minimum amount of bandwidth required for a video call Do you have high resolution video [Online]. Available: http:// support.oovoo.com/ link/portal/3908/4244/Article/1503/What-is-the-minimum-amount-of-bandwidth-required-for-a-video-call-Do-you-have-high-resolution-video. [Accessed: 03-Jun-2016].

[15] Kayange, D. S. Telemedicine Available Bandwidth Estimation Simulation Model for Effective E-Health Services: Categories, Requirements and Network Application, Acad. Res. Int., vol. 5, no. 5, pp. 1120, Sep. 2014.

[16] Tedesco, A., Di Lieto, D., Angrisani, L., Campanile, M., De Falco, M., and Di Lieto, A. Telemedicine & Broadband, Intechopen, 16-Aug-2013.

[Online]. Available: http://cdn.intechopen.com/pdfs/14281/InTech-Telemedicine_broadband.pdf. [Accessed: 03-Jun-2016].

[17] Ligthart, Leo P., and Ramjee Prasad, eds. Communications, Navigation, Sensing and Services (CONASENSE). River Publishers, 2013.

[18] Wang, Yapeng, and Ramjee Prasad. "Network Neutrality for CONASENSE Innovation Era." Role of Ict for Multi-disciplinary Applications in 2030. River Publishers, 2016. 167–184.

[19] Lala M. Purnima, Kumar, Ambuj, "Heterodox Networks An Innovative and Alternate Approach to Future Wireless Communication, Role of ICT for Multi-disciplinary Applications in 2030. Vol. 47. River Publishers, 2016.

[20] Lindgren, Peter, et al. "Open business model innovation in healthcare sector." Journal of Multi Business Model Innovation and Technology 1.1(2012):23–52.

Biographies

Ambuj Kumar received Bachelor of Engineering in Electronics & Communications from Birla Institute of Technology (BIT), Ranchi, India in the year 2000. As a part of the Bachelor programme, he carried through Internship Training in 1999 at the Institut für Hochfrequenztechnik, Technical University, (RWTH), Aachen (Germany). After graduation, Ambuj Kumar worked at Lucent Technologies Hindustan Private Limited, a vendor company, during the period 2000–2004. His major responsibilities were the Mobile Radio Network Design, Macro and Microcell planning, and Optimization for the GSM and the CDMA-based Mobile Communication Networks. There, he was involved in pan-India planning and optimizing the MCNs of various service providers for both the green field and the incumbent deployments. During the period 2004 to 2007, he worked with Hutchison Mobile Services Limited (now Vodafone), a service provider company, where he was involved in planning, deployment, and optimization of the Hutch's rapidly expanding

GSM and Edge networks across India. Afterwards, he joined Alcatel-Lucent, New Delhi in 2007 and continued until 2009. Here, he worked in Network Pre-sales Department and contributed in network rollouts for various pan-India service providers. He worked for three months in the year 2009 as Research Associate at the Centre for TeleInFrastruktur (CTIF), Department of Electronic Systems, Aalborg University, and his research area was on 'Identification of Optimization parameters for Routing in Cognitive Radio'. Ambuj Kumar was awarded research scholarship under European Commission – Erasmus Mundus "Mobility for Life" programme for doing PhD and joined CTIF, Department of Electronic Systems, Aalborg University, Aalborg, (Denmark), in the year 2010. Ambuj Kumar has also worked as a Collaborative Researcher at the Vihaan Networks Limited (VNL), India. The work of PhD research was conceptualized at VNL; there he developed testbed facilities for experimental studies on 'Advanced Alternative Networks'. He had worked as Research Assistant in the eWall Project, funded by the European Commission, at the Faculty of Science and Engineering (Department of Electronic Systems) during 2015–2016. He was awarded Doctor of Philosophy (PhD) in 2016 by the Aalborg University (Denmark) on his thesis titled "Active Probing Feedback Based Self Configurable Intelligent Distributed Antenna System for Relative and Intuitive Coverage and Capacity Predictions for Proactive Spectrum Sensing and Management". Currently, Dr. Ambuj Kumar is working as PostDoc in the Department of Business Development and Technology, School of Business & Social Sciences, Aarhus University (Denmark) since February, 2017. His research interests are radio wave propagation, cognitive radio, visible light communications, radio resource management and distribute antenna systems, etc. He has more than 10 several research publications including a book chapter in these the matic areas.

Sadia Anwar received her degree in doctor of Pharmacy in 2011 form Government College University, Pakistan. She had worked for three years as a community pharmacist and she also worked as a Clinical and Drug information consultant. She came to Denmark in December 2015 and started working as a Guest Researcher at CTiF in the department of electronic systems, Aalborg University under "IICT Endowment Fund for Sustainable Development Scheme" under the supervision of Professor Ramjee Prasad. She worked for Interdisciplinary area specifically more focused in 4 sections: medicine, Telecommunication, Big data and economics. She joined Aarhus University in 2016 and now working there as Research Assistant. Her research is based on interdisciplinary area of Medicines, ICT, and Business Development. She is also Working in vCare project (Virtual Coaching Activities for Rehabilitation in Elderly) – EU Horizon 2020 Research and Innovation funding programme.

4

Radios for Crowd Counting Applications

**Ernestina Cianca, Simone Di Domenico, Mauro De Sanctis
and Tommaso Rossi**

CGC, University of Rome Tor Vergatta, Italy

Abstract

Crowd density estimation in public spaces can be a highly effective tool for establishing global situational awareness. It is of great interest in a number of potential applications. It particular, crowd counting systems could play a key role for improving the marketing strategies, improving the customer's experience and/or improving the operational efficiency. Cost-effective solutions for these systems should be not invasive and mainly, privacy preserving. Therefore, with respect to traditional approaches that are mainly based on video-cameras, interesting approach are emerging, based on the use of radio-frequency signals of opportunity. The Chapter makes an overview of the most interesting business applications in which these systems could play an important role. Then, an overview of traditional and more novel approaches is provided. Finally, the Chapter discusses on the opportunity to bring location/context-based services, and the associated huge market opportunities, as protagonists in 5G systems.

Keywords: Crowd Counting, Operational efficiency, Privacy preserving, Radio frequency signals, Location based services.

4.1 Introduction

People counting is the process of counting or estimating the number of people in a given open area, in a crowded or in a closed environment [1]. Traditionally the motivation of crowd counting was solely for security purposes such as crowd control in public gatherings like street fairs, music festivals, theaters, stadiums. For instance, people counting can be used by security personnel and fire marshalls to estimate the number of patrons in a venue and ensure fire code compliance and guest safety.

However, we witness a growing interest for crowd counting in the business domain. As a matter of fact, crowd counting is a key part of the more general concept of physical analytics that is about getting insights on the users by understanding how he/she moves, how long they dwell, what and interactions they have (e.g., picking up items from a store shelf) in the physical worlds, exactly like online analytics tries to get the same insight from the clickstream in the online world. Understanding flow, density, direction, and activity of people allows companies and managers to gain business intelligence and insight within public parks, shopping centers, event venues, theme parks, and other public attractions. Crowd counting and pedestrian traffic monitoring are important for detecting congestion points, optimizing for visitor access, and measuring footfall traffic. Traffic counting at shopping malls or other buildings, such as museums, casinos, libraries, etc, can deliver key information to maximize potential and increase profitability.

In general, crowd counting system could be used for the following activities:

Marketing: people counting solution will provide a company with data and insights to assist in planning and evaluating marketing efforts all year long; track seasonal changes and discover the underlying trends that drive the business, so that the company can always keep one step ahead of the competition.

Set product and rent prices: by measuring the flow of traffic at all mall entrances and anchor stores, a company would be able to measure visitors in public spaces, keep track of individual store performances and adjust rent based on current data.

Optimize visitor's experience:
It would be possible to improve and ease the flow of visitors and create a more enjoyable and efficient environment by understanding: which areas are prone to congestion, which areas are busiest at different time of the day, which opportunities lay hidden in under-used parts of your building. It would be possible to adjust the light and ventilation according to the occupancy level to optimize the trade-off between user comfort and energy consumption. Such queues are often frustrating and the use of waiting and service time statistics allows managers and service providers to enhance their organization and processes to increase customer satisfaction.

Improving operational efficiency
It would be possible to improve staff scheduling around daily, weekly and seasonal variation traffic.

- Timing and frequency of maintenance and cleaning activities can be planned more efficiently.
- Security staffing can be focused more specifically on the right areas at the most appropriate time.
- Opening times and availability of various facilities can be scheduled around times of highest demand.

First of all, the Chapter focuses on one of the most representative application of crowd counting systems: the so-called shopper analytics and on the specific requirements that those systems should satisfy. Then, an overview of the more traditional and novel approaches is provided. Finally, the Chapter discusses on the opportunity to bring location/context-based services, and the associated huge market opportunities, as protagonists in 5G systems.

4.2 Shopping Analytics

Crowd counting could play a key role in the so called shopper analytics, which is the understanding of the behavior of shoppers in mall or a store, which might be indicative of their interests. As a matter of fact, unlike online shopping that can recommend related products based on customer preferences and improve the shopper's desire to purchase, the physical store cannot do in-depth analysis of shopping data. This is one of the reasons that makes online shopping have higher sales than the physical store. So, detailed and accurate physical store shopping data analysis can undoubtedly bring endless benefits to retailers and product suppliers and meanwhile provide convenience to the shoppers. In fact, the existing customer shopping data analyses during the whole shopping process merely utilize the sales history, which only reveals the hot items and related products. They cannot analyze the following facts: what kind of products are often discarded somewhere after being picked by the customers for a while, customers according to their order of selected goods, what type of clothes are always matched with or tried on together, and which area is the hot area where the manager can place an on-sale production. Those information can facilitate retailers to infer customer shopping habits, find people flow that are larger, and place promotional merchandise in that area, that is to say, optimize the shop layout, make smarter marketing strategies, like adopt bundle-selling strategies to boost profit. In the past few years, barcode is popular in commodities, but it is difficult to collect and analyze the date since the number of goods are large and collecting the date regularly can be costly and impossible [2].

4.3 Applications Requirements

In most of the mentioned applications, a crowd counting solution should meet the following criteria: i) ease of deployment - a low cost solution that leverages existing infrastructure, without requiring the installation of new sensors makes crowd counting accessible to a wider range of applications; ii) scalability - the solution should work across large geographically regions, indoors and outdoors, and scale to large number of devices for counting huge crowds; iii) Minimal intrusion - a solution that can effectively count in the background, with minimal involvement of the mobile device and user, will help to preserve anonymity, privacy and security; iv) accuracy - while a very precise count is typically not required, a fairly good estimate is necessary for most applications.

4.4 Traditional and Novel Approaches for Crowd Counting

4.4.1 Vision-based

Traditional crowd counting approaches are vision-based: images of an area are used to identify the number of people present in the area [3]. Vision based schemes require line-of-sight for monitoring and their performances degrade severely under non-ideal light conditions and situations. Moreover, they require a network of cameras to be installed in the area of interest and as such have a high deployment cost; they cannot work behind walls; usually they introduce high computational overhead and pose privacy issues.

4.4.2 Sensors-based

Crowd density estimation has been performed also by using different type of environmental sensors such temperature, concentration of carbon dioxide sensors. Also this type of solution requires installing specialized sensors and the type and granularity of information that can be gathered is rather low.

4.4.3 Device-based Solutions

The device-based or sensor-based approach aims to monitor human activities using data collected by wearable sensors, such as gyroscope, accelerometer, GPS, and other kind of inertial sensors [4]. Wearable sensors can be placed indirectly on human body, for example they can be embedded into wristbands, clothes, shoes, smart phones, or directly on it. Generally the data provided from body worn sensors are time series of physical quantities that are transmitted by a radio module to a remote server, which fuses the collected data and processes them through statistical or probabilistic model for activity recognition. The use of wearable sensors allows to overcome some of the limitations of camera-based approach. Firstly, the user privacy is preserved since no cameras are used to infer activities. Secondly, system performance are not dependent on environmental conditions, such as poor illumination or blind spots. Thirdly, since each person has its own body sensors, multi-person activity recognition is not a problem anymore. Despite these advantages, the sensor-based approach has also many drawbacks, which are mainly the following: (i) user has to wear sensors for all the monitoring time, and this requires user cooperation and availability (not possible in some applications such as surveillance); (ii) the higher the number of persons to monitor, the higher the number of required sensors, the higher the costs of the system. This means that a sensor-based approach is hardly scalable to a large number of users due to the high deployment costs; (iii) sensors are usually powered by low capacity batteries, so they have a limited life time and also require frequent battery replacements, which can represent a technical issue for real-world applications.

A subset of this class of solution is represented by the use of smarthphones or smartwatch. Many recently proposed solutions estimates the number of people in an area by counting the number of users accessing a wifi access points, or by passively monitoring the number of transmitting devices [5]. Use of smartphone is feasible and easily deployed. However, the conditions that the user carries an active transmitting device is not always verified. Moreover, all these counting approaches require data communication which may compromise anonymity, privacy and security. These communication modes leave open multiple entry points through which mobile devices could be attacked or their privacy compromised. In [6] authors propose a crowd counting solution based on audio tones, leveraging the speakerphones that are commonly available in most phones.

4.4.4 Device-free Solutions

The device-free approach does not require neither user cooperation nor wearable sensors, but it exploits electromagnetic signals, already present in the environment or generated ad hoc, to recognize activities performed by one or more persons in a given area. The key idea behind this approach is to analyze signal fluctuations caused by human movements in the target environment to infer the corresponding activity. In fact, every movement, even the smallest one, may alter the electromagnetic field in terms of diffraction, reflection and scattering, therefore by measuring that variation is possible to estimate and track human activity and its position, without requiring sensors nor user cooperation. The first studies and experiments about the use of RF signals for sensing purposes have been conducted by Youssef et. al [92] and Woyach [82] ten years ago. The device-free approach can be further classified into 2 subcategories:

- Active: the active systems employ a dedicated transmitter as part of the recognition task [7]. This means that the transmitter is under the control of the recognition system and the characteristics of the emitted signal can be properly designed (bandwidth, carrier frequency, power, etc.) to ease the recognition task. Examples of device-free active systems are those based on ultrasound waves, infrared sensors, RFID tag.
- Passive: the passive systems utilize ambient radio signals which are already present in the target environment. This means that in this case the transmitter is not under the control of the system and source of opportunity are exploited as radio transmitter, such as FM signals [8], WiFi signals [9] and Long Term Evolution (LTE) signals [10, 11].

4.5 Device-free Approaches Using Signals of Opportunity

In the framework of Internet of Things (IoT), where the "things" are embedded devices that combine functionality of sensing and communication, and where the wireless technologies, such as WiFi and LTE, are widely used, the passive device-free sensing is, of course, the most interesting approach. In fact, with the ever increasing IoT world, given an arbitrary environment is very likely to find the presence of radio signals transmitted by an already installed infrastructure and used to connect objects each other. In this scenario, a passive device-free system is extremely interesting as is able to reuse the same radio signals for the sensing purpose. In this way, with minimal additional hardware and installation costs, it is possible to extend the perception of sensing systems beyond the boundaries of an individual device or person, thus paving the way to novel services. Moreover, reusing existing RF signals the user does not change his behavior or habits, and could also not be aware of the sensing system, which is an essential feature for some applications like surveillance and security systems. Additionally to the afore mentioned advantages, the passive device-free approach allows to preserve the user privacy thanks to the use of RF signals, does not require a dedicated transmitter, and can represent a highly scalable solution. For the reasons discussed above, this approach to human sensing has attracted a significant and growing interest in the research field of human computer interaction. Since this research area is relatively recent much work has to be done and some theoretical aspects are still missing in literature.

The most important works about the use of RF signals for people counting purposes can be divided into two main categories: RSSI-based and CSI-based. As a matter of fact, most of the works are based on RSSI measurements. Xu et al. deployed an infrastructure of 20 to 22 devices to count and localize subjects in large areas [12]. The radio devices used in the experiments contain a Chipcon CC1100 radio transceiver operating in the unlicensed band at 909.1 MHz. Each transmitter periodically broadcasts a 10-byte packet every 100 ms. When the receiver receives a packet, it measures the RSS values and wraps the transmitter ID, receiver ID, RSS, timestamp (on the receiver side) into a service packet which is sent to a centralized system for data collection and analysis. Experimental results showed that SCPL works well in two different typical indoor environments of 150 m2 (office cubicles) and 400 m2 (open floor plan). In both spaces, an average accuracy of 86% accuracy for up to 4 subjects was achieved. About CSI-based crowd counting method in [13] authors applied a Gray Verhulst model to construct a normal profile of the percentage of nonzero elements of the dilated matrix of CSIs for each class. The experiments were carried out counting up to 30 persons in both indoor and outdoor settings using a single AP and 3 or 4 receivers. The results showed that this approach achieved an error ranging from 2 persons to three persons; more than 98% estimation errors were less than 2 persons in indoor environment, while about 70% errors was 2 persons in outdoor environment.

Comparison of performance and experimental setup among crowd counting methods is shown in Table 4.1. It is worth noting that all the reported methods perform a training phase in the same room where they perform the crowd counting test.

The main limitations concerning the state of the art for people counting application are the following: (i) in most of the works a high number of transmitter and/or receiver is employed, which may be a limit for system

Table 4.1 Comparison of crowd counting techniques

Ref	Environments	Max # of People	Standards	Source Data	#TXs	#RXs	Estimation Delay (Window Size W)	Crowd Counting Accuracy	Crowd Density Accuracy
[14]	2 (indoor 33 m², outdoor 70 m²)	9	WiFi	RSSI	1	1	300 s	$P(e \leq 2) = 96\%$ (outdoor) $P(e \leq 2) = 63\%$ (indoor)	–
[12]	2 (indoor m², indoor 400 m²)	4	ISM 909.1 MHz	RSSI	12/13	8/9	1 s	84%	–
[12]	2 (indoor 150 m², indoor 400 m²)	4	ISM 909.1 MHz	RSSI	12/13	8/9	1 s	84%	–
[15]	1 (indoor 324 m²)	>10	ISM 2.4 GHz	RSSI	16	16	–	–	86% (op-3p, 4p-10p, >10p)
[16]	1 (indoor 100 m²)	15	ZigBee 2.4 GHz	RSSI	1	3	–	–	73% (5p, 10p, 15p)
[17]	1 (indoor)	7	WiFi	RSSI	1	10	300 s	77%	94% (op, 1p-3p, 4p-7p)
[18]	1 (indoor corridor ≈ 10 m²)	5	HBE-Zigbex	RSSI	1	2	4.5 s	77%	–
[19]	2 (indoor, outdoor)	30	WiFi	CSI	1	3/4	–	$P(e \leq 2) = 70\%$ (outdoor) $P(e \leq 2) = 98\%$ (indoor)	–

deployability and installation in practical scenarios. (ii) all the above-mentioned works use a training-based approach, i.e. the system is trained and tested in the same environment under the same configuration. This means that the system calibration needs to be performed again for each new environment to test.

Novel solutions have been presented recently [20, 21], which allow the installation of a low number of devices and are able to reduce in the need of training. The idea is to the use of a differential approach where the information on the number of people is inferred by how the channel frequency response changes over time and not from its absolute value which depends also from the background environment. A similar approach has been used in [10] where the signal of opportunity is LTE. In [10], authors showed through experimental results that the LTE signal can be exploited to provide a rough estimation of the number of people inside the room where the LTE receiver is placed. The average recognition accuracy ranged from 72% to 95% and was over three different experimental setups in the same room. It is worth noting that the empty room was recognized by the system with an accuracy of 100% and none of the considered classes of people was misclassified with the empty room. Obtained results were very promising, also compared with the results achieved by other literature crowd counting systems using WiFi signals. Moreover, the proposed system was also tested in through the wall conditions, i.e. by placing the LTE receiver in the adjacent room to the one in which the people were moving. In this case, the system was able to distinguish the empty room, the presence of one person, and the presence of multiple persons, with an average accuracy of about 90%. The through the wall detection can be very useful for many security applications in which the target environment cannot be directly controlled by the security or military guards. The achieved results are very promising and demonstrate the feasibility of an RF based device-free crowd size estimation system based on LTE signals. It is worth outlining that this is the first in literature that attempts to use the LTE signal for people counting purposes, so it paves the way to new challenges and future research works.

4.6 Future Perspectives and Conclusions

First of all, it must be outlined that crowd counting could system could play an important role in the framework of context-aware communications, where sensing the context could also facilitate the network management and protocol adaptation. Knowing about the number of users in an area, IoT systems can be made truly smart and adaptable to the context they operate in, and not just operate based on a predefined set of rules or threshold values [5].

Moreover, could the development of 5G takes into account the importance of context-awareness? In other words, would it be possible to think the 5G architecture where localization/context, together with analytics, are combined and provided "as a service"? This would greatly increase the overall value of the 5G ecosystem, allowing network operators to better manage their networks and to dramatically expand the range of offered applications and services.

In this perspectives, 5G provides interesting opportunities related to the use of millimeters waves, which are supposed to be use by 5G networks for communication purposes (very high data rate short-range communications). However, millimeter wave radios are also known to be good sensors [22]. Therefore, their use could enable standalone mobile sensing capabilities, as proposed in [23]. Moreover, this ability to sense the context, such as position/orientation and density of human activities, could be used for optimizing the network management or facilitating the node deployment. An preliminary study on an antenna beam adaptation based on the context is investigated in [24].

To summarize, the ongoing technological evolution in 5G designs is already providing several key assets towards a integration of communication and sensing and their combined provision "as a service". (such as ultra-dense small cell coverage, advanced waveforms, access convergence and integration of heterogeneous technologies). These assets, if properly exploited and adapted since the very beginning of the ongoing 5G network design, can provide the opportunity to bring location/context-based services, and the associated huge market opportunities, as protagonists in 5G systems. However, research activities are still fragmented and there is no common effort to natively incorporate localization and sensing capabilities in 5G and make of the 5G network infrastructure, an infrastructure truly and simply usable by third party stakeholders to develop their services.

References

[1] E. Cianca, M. De Sanctis, and S. Di Domenico. "Radios as Sensors." In: IEEE Internet of Things Journal 4.2 (2017), pp. 363–373. issn: 2327-4662. doi: 10.1109/JIOT.2016.2563399

[2] Jumin Zhao, Like Wang, Deng-ao Li, Yanxia Li, Bin Yang, Biaokai Zhu and Ruiqin Bai, "Mining shopping data with passive tags via velocity analysis", *EURASIP Journal on Wireless Communications and Networking* (2018) 2018:28 DOI 10.1186/s13638-018-1033-5

[3] Shian-Ru Ke, Hoang Le Uyen Thuc, Yong-Jin Lee, Jenq-Neng Hwang, Jang-Hee Yoo, and Kyoung-Ho Choi. "A Review on Video-Based Human Activity Recognition". In: Computers 2.2 (2013), pp. 88–131. issn: 2073-431X. doi: 10.3390/computers2020088.

[4] L. Chen, J. Hoey, C. D. Nugent, D. J. Cook, and Z. Yu. "Sensor-Based Activity Recognition". In: IEEE Transactions on Systems, Man, and Cybernetics, Part C (Applications and Reviews) 42.6 (2012), pp. 790–808. issn: 1094-6977. doi: 10.1109/TSMCC. 2012.2198883.

[5] Lars Møller Mikkelsen, "Enhancing IoT Systems by Exploiting Opportunistically Collected Information from Communication Networks", Ph.D. Dissertation, 2017, Aalborg University.

[6] P. G. Kannan, S. P. Venkatagiri, M. C. Chan, A. L. Ananda, and L. Peh, "Low cost crowd counting using audio tones," Proc. 10th ACM Conf. Embed. Netw. Sens. Syst. - SenSys '12, p. 155, 2012.

[7] Fadel Adib, Zachary Kabelac, Dina Katabi, and Robert C. Miller. "3D Tracking via Body Radio Reflections." In: Proceedings of the 11th USENIX Conference on Networked Systems Design and Implementation. NSDI'14. Seattle, WA: USENIX Association, 2014, pp. 317–329. isbn: 978-1-931971-09-6.

[8] S. Sigg, M. Scholz, S. Shi, Y. Ji, and M. Beigl. "RF-Sensing of Activities from Non-Cooperative Subjects in Device-Free Recognition Systems Using Ambient and Local Signals". In: IEEE Transactions on Mobile Computing 13.4 (2014), pp. 907–920. issn: 1536–1233. doi: 10.1109/TMC.2013.28.

[9] Yan Wang, Jian Liu, Yingying Chen, Marco Gruteser, Jie Yang, and Hongbo Liu. "E-eyes: Device-free Location-oriented Activity Identification Using Fine-grained WiFi Signatures". In: Proceedings of the 20th Annual International Conference on Mobile Computing and Networking. MobiCom '14. Maui, Hawaii, USA: ACM, 2014, pp. 617–628. isbn: 978-1-4503-2783-1. doi: 10.1145/2639108.2639143.

[10] S. Di Domenico, M. De Sanctis, E. Cianca, P. Colucci, and G. Bianchi. "LTE-based passive device-free crowd density estimation." In: 2017 IEEE International Conference on Communications (ICC). 2017, pp. 1–6. doi: 10.1109/ICC.2017.7997194.

[11] G. Pecoraro, S. Di Domenico, E. Cianca, and M. De Sanctis. "LTE signal fingerprinting localization based on CSI." In: 2017 IEEE 13th International Conference on Wireless and Mobile Computing, Networking and Communications (WiMob). 2017, pp. 1–8. doi: 10.1109/WiMOB.2017.8115803.

[12] Chenren Xu, Bernhard Firner, Robert S. Moore, Yanyong Zhang, Wade Trappe, Richard Howard, Feixiong Zhang, and Ning An. "SCPL: Indoor Device-free Multisubject Counting and Localization Using Radio Signal Strength." In: Proceedings of the *12*th International Conference on Information Processing in Sensor Networks. Philadelphia, Pennsylvania, USA: ACM, 2013, pp. 79–90. doi: 10.1145/2461381.2461394.

[13] Wei Xi, Jizhong Zhao, Xiang-Yang Li, Kun Zhao, Shaojie Tang, Xue Liu, and Zhiping Jiang. "Electronic frog eye: Counting crowd using WiFi." In: *2014* IEEE Conference on Computer Communications, INFOCOM *2014*, Toronto, Canada. 2014, pp. 361–369.

[14] S. Depatla, A. Muralidharan, and Y. Mostofi. "Occupancy Estimation Using Only WiFi Power Measurements." In: Selected Areas in Communications, IEEE Journal on 33.7 (2015), pp. 1381–1393.

[15] Yaoxuan Yuan, Jizhong Zhao, Chen Qiu, and Wei Xi. "Estimating Crowd Density in an RF-Based Dynamic Environment." In: Sensors Journal, IEEE 13.10 (2013), pp. 3837–3845.

[16] Solahuddin Yusuf Fadhlullah and Widad Ismail. "A Statistical Approach in Designing an RF-Based Human Crowd Density Estimation System". In: International Journal of Distributed Sensor Networks 2016 (2016), 8351017:1–8351017:9. doi: 10.1155/2016/8351017.

[17] Takuya Yoshida and Yoshiaki Taniguchi. "Estimating the number of people using existing WiFi access point in indoor environment". In: Proceedings of the 6th European Conference of Computer Science (ECCS '15). Rome, Italy, 2015, pp. 46–53.

[18] S. H. Doong. "Spectral Human Flow Counting with RSSI in Wireless Sensor Networks". In: 2016 International Conference on Distributed Computing in Sensor Systems (DCOSS). 2016, pp. 110–112. doi: 10.1109/DCOSS.2016.33.

[19] Wei Xi, Jizhong Zhao, Xiang-Yang Li, Kun Zhao, Shaojie Tang, Xue Liu, and Zhiping Jiang. "Electronic frog eye: Counting crowd using WiFi". In: 2014 IEEE Conference on Computer Communications, INFOCOM 2014, Toronto, Canada. 2014, pp. 361– 369.

[20] Simone Di Domenico, Mauro De Sanctis, Ernestina Cianca, and Giuseppe Bianchi. "A Trained-once Crowd Counting Method Using

Differential WiFi Channel State Information." In: Proceedings of the 3rd International onWorkshop on Physical Analytics. WPA '16. Singapore, Singapore: ACM, 2016, pp. 37–42. isbn: 978-1-4503-4328-2. doi: 10.1145/2935651.2935657.

[21] S. Di Domenico, G. Pecoraro, E. Cianca, and M. De Sanctis. "Trained-once devicefree crowd counting and occupancy estimation using WiFi: A Doppler spectrum based approach". In: *2016* IEEE 12th International Conference on Wireless and Mobile Computing, Networking and Communications (WiMob). 2016, pp. 1–8. doi: 10.1109/WiMOB.2016.7763227.

[22] Xinyu Zhang, "Millimeter-wave for 5G: Unifying Communication and Sensing", Microsoft Research Faculty Summit 2015.

[23] Teng Wei, Xinyu Zhang, "m Track: High Precision Passive Tracking Using Millimeter-wave Radios", ACM MobiCom'15.

[24] S. Sur, V. Venkateswaran, X. Zhang and P. Ramanathan, "60 GHz Indoor Networking through Flexible Beams: A Link-Level Profiling", ACM SIGMETRICS'15.

Biographies

Ernestina Cianca is Assistant Professor at the Dept. of Electronic Engineering of the University of Rome Tor Vergata, where she teaches Digital Communications and ICT Infrastructure and Applications (WSN, Smart Grid, ITS etc.). She is the Director of the II Level Master in Engineering and International Space Law in Satellite systems for Communication, Navigation and Sensing. She is vice-director of the interdepartmental Center CTIF-Italy. She is the Editor-in Chiefs of the CONASENSE (Comm/Nav/Sensing and services) Journal, River Publishers.

She has worked on wireless access technologies (CDMA, OFDM) and in particular in the waveforms design, optimization and performance analysis of radio interfaces both for terrestrial and satellite communications. An important part of her research has focused on the use of EHF bands (Q/V band, W band) for satellite communications and on the integration of satellite/terrestrial/HAP (High altitude Platforms) systems. Currently her main research interests are in the use of radio-frequency signals (opportunistic signals such as WiFi or specifically designed signals) for sensing purposes, and in particular. Device-free RF-based activity recognition/crowd counting/density estimation and localization; UWB radar imaging (i.e., stroke detection). She is author/co-author of 100 papers in international journals and conferences.

Simone Di Domenico received both Bachelor and Master degrees in Internet technology engineering from the University of Roma "Tor Vergata", in 2012 and 2014, respectively. Currently, he is a Ph.D. student in Electronic Engineering at the University of Roma "Tor Vergata" and his main research interests include the RF device-free human activity recognition and the RF device-free people counting.

Mauro De Sanctis received the Ph.D. degree in Telecommunications and Microelectronics Engineering in 2006 from the University of Roma "Tor Vergata" (Italy). From the end of 2008 he is Assistant Professor in the

Department of Electronics Engineering, University of Roma "Tor Vergata" (Italy), teaching "Information Theory and Data Mining". From January 2004 to June 2008 he has been involved in the MAGNET European FP6 integrated project as scientific responsible of the activities on radio resource management. He has been involved in research activities for several projects funded by the Italian Space Agency and in several Italian Research Programs of Relevant National Interest. He is serving as Associate Editor of the IEEE Aerospace and Electronic Systems Magazine. His main areas of interest include: wireless terrestrial and satellite communication networks, data mining and information theory.

Tommaso Rossi received his University Degree in Telecommunications in 2002, MSc Degree in Advanced Communications and Navigation Satellite Systems in 2004 and PhD in Telecommunications and Microelectronics in 2008 at the University of Rome Tor Vergata where he is currently an Assistant Professor (teaching Digital Signal Processing, Multimedia Processing and Communication and Signals). His research activity is focused on Space Systems, EHF Satellite and Terrestrial Telecommunications, Satellite and Inertial Navigation Systems, Digital Signal Processing for Radar and TLC applications. He is currently Co-Investigator of the Italian Space Agency Q/V-band satellite communication experimental campaign realized through the Alphasat Aldo Paraboni payload. He is Associate Editor for the Space Systems area of the IEEE Transactions on Aerospace and Electronic Systems.

5

Applications of CONASENSE

Sriganesh K. Rao and Ramjee Prasad

CGC, Aarhus University, Denmark

Abstract

The vision of CONASENSE is – To Integrate the COmmunications, NAvigation, SENsing and SErvices. It is a common platform aimed at synergistic combination of technologies for communications, positioning and sensing systems. The goal of the "integrated vision" is to improve the user's Quality of life.

This chapter describes the role of Satellites and Unmanned Air Vehicles (UAVs) in implementing the integrated vision of CONASENSE. It lists out various CONASENSE applications enabled by Satellites and UAVs.

Keywords: CONASENSE, Satellites, Unmanned Air Vehicles, Positioning, Sensing.

5.1 Introduction

The Vision of CONASENSE is to integrate the **CO**mmunications, **NA**vigation, **SEN**sing and **SE**rvices. The goal of the "Integrated Vision" is to improve the user's Quality of Life.

Communication, Navigation and Sensing systems can mutually assist each other by exploiting a bi-directional interaction amongst them, to give rise to new business Services opportunities. It is a common platform aimed at synergistic combination of technologies for communications, positioning and sensing systems able to effectively use information about location, context and situation of several entities [1].

'**Location**' – Refers to geographical coordinates of mobile users. It can be extended by including other information like speed, acceleration, direction and orientation.

'**Context**' – Describes the environment in which the user is embedded and the devices/access networks with which the user interacts. User context can consist of attributes such as physiological/emotional/activity states and environmental context includes light, sound, weather, humidity, temperature, noise etc.

'**Situation**' – Refers to the interpretation of the contextual information that can be related to a user or access network.

''**Context**' – is an objective description of the environment while "Situation" is a subjective interpretation of a Context.

Satellites and Unmanned Air Vehicles are two important factors in CONASENSE.

5.2 Role of Satellites in CONASENSE

Satellites will integrate with other networks rather than be a stand-alone network and it is this integration that forms the core of the vision of CONASENSE. Satellite systems are fundamental components to deliver reliably CONASENSE services across the world, all the time and at an affordable cost. The satellite component will contribute to augment the service capability and address some of the major challenges in relation to the support of multimedia traffic growth, ubiquitous coverage, machine to machine communications and critical telecom missions whilst optimising the value for money to the end-users.

Satcom systems can address a wide range of services such as broadcast, broadband and narrowband services to fixed, portable and mobile terminals over global or regional coverage [2]:

- Broadcast systems have been optimized to deliver TV programs.
- Broadband services support IP services

Global coverage and dependability are the main added value of space-based communication services. Integrated in telecom network infrastructure, SatCom solutions are well positioned to target following types of use cases [3]:

1. Multimedia Distribution: Estimated that 90% of all Internet traffic is video. In order to offload the traffic, the simultaneous use of satellite

broadcast for linear content (such as video, including UHD and HD video) combined with the use of broadband networks for the non-linear content, is the most cost and spectrum efficient means of transmitting audio-visual linear services including HDTV, 4K, both feeding local distribution networks with content and delivering this content directly to end-users.

2. Service Continuity: With its regional or worldwide coverage, SatCom solutions are essential to provide the communication service everywhere including remote areas, on board vessels, aircraft (In-flight services) and trains in a reliable manner. Satellite systems can contribute to extend the coverage either providing backhaul or direct access service.

3. M2M: The inclusion of billions of sensors and actuators all transmitting low date rates and being scattered over wide and remote areas makes it well suited to data collection and control via satellite. In particular, monitoring/surveillance of various assets (vehicles, homes, machines, etc.) in remote locations, asset tracking (e.g. container) and transfer data and/or configuration to a group of widespread recipients requires satellite systems to ensure service continuity.

The key areas that satellites can contribute in CONASENSE

- Extending the coverage of communication networks.
- Delivering multimedia closer to the edge to improve latency.
- Off-loading the terrestrial network e.g. by using satellite backhaul and by taking control traffic via satellite.
- Providing resilience by integrating satellite with terrestrial.
- Integrating virtualisation via software defined networking.
- Allowing improved utilisation of the spectrum between the systems.
- Satellites will play an important role in the extension of cellular networks to sea, air and remote land areas. With many more people expecting to have the same coverage when travelling (on cruise liners, passenger aircraft, high speed trains and in holiday villas) it is key that satellite allows seamless extension of communication services.
- IoT coverage to wide areas involving sensors and M2M connections are ideal services to make use of satellite wide area coverage. The challenge is to design efficient low data rate communications in large numbers via the satellite.
- Transport services including Vehicle-to-Vehicle are again ideal for satellite with its wide coverage. In the safety market all new vehicles are likely to be mandated to include safety features.

- Satellites are already used for earth resource data which is in itself used as an input to many new services. Coupling this with integrated satellite & cellular communications will provide a powerful new fusion enabling the innovation of services.

5.3 Role of Unmanned Air Vehicles (Drones) in CONASENSE

'Unmanned Air Vehicles' or 'Drones' forms a key player in delivering CONASENSE services.

Drones represent the integrated vision of communications, navigations, sensing and services. A Drone application consist of a drone, payload and data processing [4].

A UAV is an aircraft without a human pilot aboard. UAVs are a component of an Unmanned Aircraft System (UAS) which includes UAV, a ground-based controller and a system communication between the two. The flight of UAVs may operate with various degrees of autonomy – either under remote control by human operator or autonomously by onboard computers. They were originally used for missions which were dirty or dangerous to humans and were mainly confined to military applications. However, now their use is expanding to commercial, scientific, recreational, agricultural and other applications such as policing, peacekeeping, surveillance, product deliveries, aerial photography, agriculture etc.

A UAV is defined as a "powered, aerial vehicle that does not carry a human operator, uses aerodynamic forces to provide vehicle lift, can fly autonomously or be piloted remotely, can be expendable or recoverable, and can carry a lethal or nonlethal payload". Therefore, missiles are not considered UAVs because the vehicle itself is a weapon that is not reused, though it is also unmanned and, in some cases, remotely guided [5].

UAVs typically fall into one of six functional categories:

- Target and decoy – providing ground and aerial gunnery a target that simulates an enemy aircraft or missile
- Reconnaissance – providing battlefield intelligence
- Combat – providing attack capability for high-risk missions
- Logistics – delivering cargo

- Research and development – improve UAV technologies
- Civil and commercial UAVs – agriculture, aerial photography, data collection

Drones are classified as Fixed Wings, Multi Rotors and Hybrids

5.3.1 Fixed Wing Drones

Figure 5.1   Example 1 of Fixed Wing Drone.

Figure 5.2   Example 2 of Fixed Wing Drone.

Figure 5.3   Example 3 of Fixed Wing Drone.

5.3.2 Multi Rotor Drones Are Shown Below

Figure 5.4   Example 1 of Multi Rotor Drone.

Figure 5.5   Example 2 of Multi Rotor Drone.

Figure 5.6   Example 3 of Multi Rotor Drone.

5.3.3 Hybrid Drones

Figure 5.7 Example of Hybrid Drone.

5.4 Applications of CONASENSE Based on Satellites

5.4.1 Disaster Monitoring and Real-Time Management

Large scale disasters like floods, earthquakes, forest fires, Tsunami etc. — processed Earth Observation data can be used to predict in real time, the evolution of the situation to enable disaster management resources to be pre-positioned in high risk areas.

Figure 5.8 Monitoring landslides created disasters.

Figure 5.9   Monitoring flooding created disasters.

5.4.2 Telemedicine for eHealth

The local Sensor network measuring patient parameters, is integrated into a Wider network for remote monitoring/supervision by a medical center.

5.4.3 Mountain Rescue

Multiple search & rescue teams in a very large (including unexplored) area, has to be coordinated through integrated communication and satellite navigation system.

5.4.4 Maritime Surveillance

Maritime surveillance involves integration of Communications, Navigation and Earth Observatory systems.

5.4.5 Road Traffic Optimization

Earth observation systems provides images and information related to road conditions, in terms of traffic or road accidents, in real-time. These data can be integrated with local network and every single user can be provided data in real time about road conditions and optimal routes.

5.4.6 Oil Spill

Oil & Gas companies operate very long pipeline network along remote and hostile terrains. Integration of satellite sensors can give real-time updates on the status of the pipelines, to enable prevention or very fast intervention in case of leakage from pipelines.

5.4.7 Precision Agriculture/Farming

Involves application of positioning and remote sensing technologies to large farms, to automatically regulate the use of fertilizers, pesticides and water through automatic guidance of farm vehicles.

5.4.8 Transportation – Fleet Management & Cargo Condition Monitoring

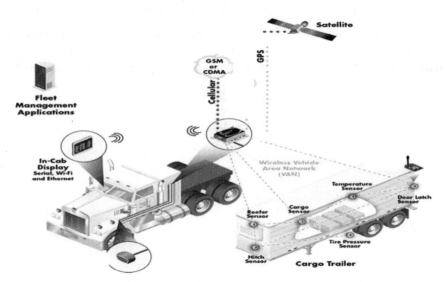

Figure 5.10   Satellite based fleet management.

5.4.9 Emergency Call [6]

eCall (emergency call) is a regulated service mandating all new vehicles to be connected to mobile communication networks and to be capable of geo-location by means of European Global Navigation Satellite System (E-GNSS/Galileo) receivers. Thus, eCall marks the beginning of the adoption of connected services on a larger scale.

An emergency call (eCall) is made automatically by the car as soon as on-board sensors (eg airbag sensors) registers a serious accident. By pushing a dedicated button in the car, any car occupant can also make an eCall manually.

The accurate position of the accident scene is fixed via satellite positioning and mobile caller location and this is transmitted to the nearest emergency call centre. At the Emergency call center, the eCall's urgency is recognized and the accident location can be viewed and emergency services can be sent off immediately. Due to the exact knowledge of the accident's location, emergency services (ambulances, fire engine, police etc) arrive at the crash site at the earliest possible instant, saving lives.

5.4.10 Vehicle to Vehicle (V2V) and Vehicle to Infrastructure (V2I)

V2V communications comprises a wireless network where automobiles send messages to each other with information about what they're doing. This data would include speed, location, direction of travel, braking, and loss of stability.

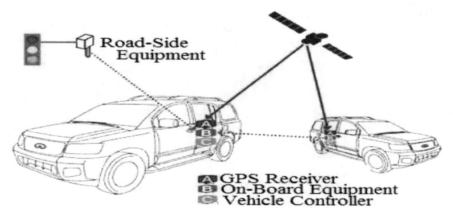

Figure 5.11   Satellite based V2V and V2I.

5.4.11 Autonomous Vehicles

A fully autonomous self-driving vehicle will need to perform overtake maneuvers on both unidirectional and bi-directional roads, where oncoming vehicles may be well beyond the range of its sensors, but approaching very quickly. Performing such maneuvers safely will require cooperation among vehicles on multiple lanes, to create the necessary gap to allow the overtaking vehicle to quickly merge onto the lane corresponding to its direction of travel in time to avoid a collision with an oncoming vehicle. This is a very good application of CONASENSE.

5.5 Applications of CONASENSE Based on UAVs [7]

5.5.1 Inspection and Monitoring

- On- & offshore inspection of oil and gas platforms.
- Inspection of infrastructure like bridges, roads, railways, power lines & equipment, viaducts, subways, tunnels, level crossings, dams, reservoirs, retaining walls etc.
- Inspection of wind farm & power stations.
- Inspection of solar park & PV modules.
- Thermal energy efficiency inspections.
- Inspection of drilling rigs, pipelines & transmission network.
- Asset & utility inspection.
- Maintenance Surveys – Access areas that are not normally reachable.
- Rooftop Inspections/Surveys (Residential Heat and Leaks)
- Industrial Roof Inspections (Heat and Leaks)

5.5.2 Videography/Photography

- Home movies
- Kids sporting events
- Weddings
- Golf Course
- Promotional Videos for Products/Services
- 360 Panoramas
- Tourism Aerial landmark flyby
- News
- Sports Event Coverage
- Live Event Coverage
- Concerts
- Traffic Reporting
- Aerial cinematography for movies
- Action/sports
- Documentary/Expedition

5.5.3 Real Estate

- Real estate/property photos and video
- Residential real estate marketing
- Commercial real estate marketing
- 360 Area shots

5.5.4 Disaster Response

- Search and rescue
- Marine search and rescue
- Wildfire
- Flooding
- Damage Assessment
- Rapid response
- Surf lifesaving
- Fire detection

5.5.5 Government

- Police surveillance
- Border Patrol/Detection
- Drug Enforcement
- Grow-op Detection
- Forestry and Fire Protection
- Highway Patrol and Enforcement
- Bridge Inspection
- Asset Management
- Military Uses
- Weather atmospheric studies
- Coastguard
- Poaching for fisheries

5.5.6 Environment

- Land cover mapping
- Carbon capping
- Renewable energy
- Environmental Monitoring (dumping)
- Waterway Monitoring

- Ice Flow Monitoring
- Parks (Asset management)
- Wildlife Conservation
- Conservation drones
- Wildlife counts/Mapping of animal population
- Anti-Poaching
- Marine Biology

5.5.7 Agriculture

- Irrigated land mapping
- Crop type
- Plant count
- Canopy cover
- Leaf area index
- Soil type and classification
- Soil moisture
- Growth stage
- Plant Height
- Nitrogen deficiencies
- Plant health
- Yield Monitoring
- Cattle herding
- Spectral imaging

5.5.8 Tourism

- Remote tours of distant locations
- Interesting landmarks

5.5.9 Utilities/Mining/Oil/Gas

- Radio Tower Inspections
- Hydro-line/Power-Line Inspection
- Wind Turbine Inspections
- Vegetation management
- Asset verification
- Solar Energy Site Assessment

- Solar Panel Outage Detection
- Oil spill tracking
- Pipeline monitoring
- Environmental assessment
- Pit survey

5.5.10 Mapping

- Land cover mapping
- Forestry mapping
- Biomass
- Forest health
- Disease detection
- Environmental mapping
- Hydrology & Geological Mapping
- Water management support mapping
- Wind Farm Mapping
- Solar power plant mapping
- Transmission Line mapping
- Emergency response mapping
- Disaster Site Monitoring and mapping
- Hazard Mapping
- Archaeological Site Mapping
- Tree Mapping

5.6 Conclusions

Satellites and Unmanned Air Vehicles play a major role in implementing the integration vision of CONASENSE. A lot of important services are enabled by them. Satellites enabled services include extending the coverage of communication networks, delivering multimedia closer to the edge. Off-loading terrestrial network, IoT coverage to wide areas involving sensors and M2M connections.

The applications enabled by UAVs are expanding to commercial, scientific, recreational, agricultural and other applications such as policing, peacekeeping, surveillance, product deliveries, aerial photography etc.

References

[1] www.conasense.org
[2] Networld2020's – SatCom WG ver 5–31st July 2014.
[3] Networld2020's – SatCom WG ver 5–31st July 2014.
[4] "Applications of UAVs" by Jakob Jakobsen, National Space Institute 2016.
[5] https://en.wikipedia.org/wiki/Unmanned_aerial_vehicle
[6] GSMA mAutomotive "2025 Every Car Connected" Feb 2012 Ver 1.0
[7] Product catalogues of Ascending Technologies and AirVid Inc.

Biography

Sriganesh Rao received his B.E (Electronics) degree from Bangalore University (India) in 1980 and MTech (Electronics) degree from NITK, Surathkal (India) in 1983 followed by MBA (Technology Management) of Deakin University, Australia in 2000 with focus on International Telecommunication Management. He is pursuing PhD at Department of Business and Technology, Aarhus University, Denmark.

He is an ICT Professional with over 34 years of global experience in leading complex technology projects in Australia and India, with rich background in Business Development, Engineering, Program Management and QA in Telecom, Government & Defence verticals across Embedded, Engineering Services, Testing and Manufacturing horizontals.

His experience spans across leading IT organizations like Tata Consultancy Services & IBM, Telecom Service Providers like TATA Teleservices & C&W OPTUS (Australia), Telecom OEMs like Bharat Electronics, United Telecoms & ERG (Australia) and Telecom OSS/BSS Providers like Alopa Networks. At present, he is with Tata Consultancy Services, Bangalore (India) working in Telecom domain. He has handled Business development and Pre-sales Solutioning in Telecom, Media & Entertainment, Government, Defence & Aerospace market segments and also led strategic initiatives in new technology areas–IPv6 and M2M/IoT for India business.

He is a Senior Member of IEEE, Senior Member of Australian Computer Society and Life Member of Computer Society of India. He is actively involved in M2M/IoT related initiatives of Professional bodies like IEEE, GISFI, TM Forum, ACS and TiE. As the Chairman for the Dept. of Telecom's (Govt of India) "M2M Gateway and Architecture" Workgroup and Member of its Joint Editorial Team on M2M Enablement in Remote Health Management, Power Sector, Intelligent Transport Systems and Safety & Surveillance Systems, he has contributed to the preparation of "National Telecom M2M Roadmap". He continues as an active Member of the National Working Group of DoT's "IoT and its applications in Smart Cities".

His current interests include 5G, M2M, IoT, Smart Cities, Industries 4.0 and Business Model Innovation.

6

Cooperative Wireless Sensor Networks: A Game Theoretic Approach

Homayoun Nikookar and Herman Monsuur

Netherlands Defence Academy

Abstract

Wireless sensor network (WSN) is a network of low-size and low-complexity devices that sense the environment and communicate the gathered data through wireless channels. The sensors sense the environment, and send data to control unit for processing and decisions. The data is forwarded via multiple hops or is connected to other networks through a gateway. WSNs have a wide range of applications from monitoring environment and surveillance, to precision agriculture, and from biomedical to structural and infrastructure health monitoring. Technological advances in the past decades have resulted in small, inexpensive and powerful sensors with embedded processing and radio networking capability. Distributed cooperative smart sensor devices networked through the radio link and deployed in large numbers provide enormous opportunities. Cooperation between nodes will increase the performance of wireless sensor network in fulfilling its task. Game theory is used to describe how and why coalitions of sensor nodes form, using the trade-off between the advantage of cooperation (in terms of better performance) and the costs of cooperation (in terms of bandwidth, transmitting information), using the topology of the network. A general game-theoretic framework for WSN is presented, and illustrated by means of an example.

Keywords: Traffic control, Internet, Game theory, Surveillance, Sensor network.

Towards Future Technologies for Business Ecosystem Innovation, 93–108.

6.1 Introduction

Wireless Sensor Network is a network of (micro)sensors which are low-size and low-complexity devices (nodes). These nodes sense the environment and communicate the information gathered from the monitored field through wireless links. The gathered data is forwarded, possibly via multiple hops relaying, to a sink that can use it locally, or is connected to other networks (e.g., the Internet) through a gateway. Wireless sensor networks improve the performance of detection and observation of environment through the employment of geometric diversity. Development of sensor networks requires three technologies: Sensing, Communication and Computing [1]. One major example of early sensor networks is the Air Traffic Control. However, the main driving force for the early application of (wireless) sensors was defence applications of military sensors. During World War II acoustic sensors (hydrophones) were developed in the oceans bottom to detect Soviet submarines. In the Cold War era networks of air defense radar were deployed including AWACS planes which was followed by distributed sensor networks research for DARPA (Defense Advanced Research Projects Agency) in 1980 using ARPANET (which was in fact the predecessor of Internet). Advances in MEMS (Micro-electro-mechanical system) technology, wireless networking, and inexpensive low-power processors are important factors for massive development of wireless sensor networks. Sensor networks in the 21st century are, among others, ad hoc, especially for highly dynamic environments, and have network information processing capability. The WSN Technology trends perspective entails advances in wireless networks, advances in chip capacity and processor production (causing energy/bit reduction for computing and computation), and sensing, computing and communication integrated on a single chip resulting in the cost reduction as well as the deployment of sensors in large numbers.

A typical sensor node in a WSN has a small size (at the size of a button) which accommodates sensors, radio transceivers, a small processor, a memory and a power unit See Figure 6.1.

With the proliferation of wireless sensor networks the requirements on prime resources like battery power and radio spectrum are put under severe pressure. In a wireless environment the system requirements, network capabilities and device capabilities have enormous variations giving rise to significant design challenges. There is therefore an emergent need for developing energy efficient, green technologies that optimize premium radio resources, such as power and spectrum, even while guaranteeing quality of

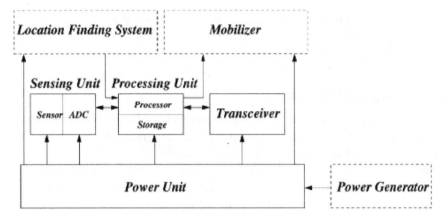

Figure 6.1   Sensor Node Architecture [2].

service (for example the required security). Moreover, many wireless sensor networks working in monitoring and surveillance scenarios operate under dynamic conditions with frequent changes in the propagation environment and diversified requirements. All these trends point to flexible, reconfigurable structures that can adapt to the circumstances and the radio neighbourhood and can also cooperate with other sensor networks. The nodes of future WSNs will most likely be context aware, cooperative and have energy harvesting capabilities.

6.2 Challenges in WSN

The challenges of WSNs, in general, are manifold. Topology of sensor network can be seen as the first challenge. Each node needs to know the identity and location of neighbours in the fixed or mobile scenarios. Furthermore, the planned networks a-priori knowledge is needed. Furthermore, for the adhoc sensor networks the topology should be made in the real time and the updated- failed nodes should be out and the new sensors be included in the network.

The second challenge is the network control and routing. Dealing with the resources-energy, bandwidth and processing power, Protocols, Mobile IP (needing heavy overhead-time, memory, energy), size and density of network, and trade-off between latency, reliability and energy are further challenges to name.

In addition to these challenges, and in particular for the WSN applications in monitoring scenarios, further challenges are envisaged. In the following these particular issues are discussed in more detail.

6.2.1 Propagation Channel

In wireless sensor networks typically the nodes have low-height antennas. In these applications the radio propagation channel characteristics and among others the path loss exponent is considerably different from the free-space channel. Therefore, routes with more hops and with shorter hop distances can be more power-efficient than those with fewer hops but longer hop distances. One of the major research challenges in this regard is to investigate the impact of dynamic channel on the adaptive self-configuration topology mechanism of sensor network and see to which extent it can reduce energy consumption and increase network performance.

6.2.2 Cognitive WSNs

Current WSNs operate in the ISM (Industrial-Scientific-Medical) band. This band is shared by many other wireless technologies giving rise to degradation of performance of WSNs due to interference. WSNs can also interfere other services in this band. The proliferation of WSNs will result in scarcity of spectrum dedicated to wireless sensor communications. Cognitive Radio (CR) technology for WSNs improves sensor nodes communications performance as well as spectral efficiency. It is foreseen that cognitive radio will emerge as an active research area for wireless networks in the coming years. Unlike conventional radios in which most of components are implemented in hardware, cognitive radio uses software implementations (i.e., Software Defined Radio (SDR)[1]) for some functionalities enabling a flexible radio operation. The radio sensors are reconfigurable and therefore the need to modify existing hardware is reduced. In this context the increasing number of sensing nodes equipped with wireless communication capability will require faster connectivity and thus, wireless spectrum will have to be adapted

---

[1] Software Defined Radio is a software based, programmable and reconfigurable modulation and demodulation technique. With the flexibility that it provides, hybrid platforms can be deployed in the wireless sensor network. By integrating SDR technique in the WSN, with the same (programmable) hardware, more radio standards can be introduced to the network. Therefore, instead of designing again the hardware, only sensor nodes of the WSN are reprogrammed.

to the new requirements (of bandwidth). Cognitive radio will prevent the need to implement hardware upgrades. It will allow the cognitive radio-enabled sensor nodes to search for the best frequency-based pre-determined parameters. It should be noted that CR-WSNs differ from the conventional WSNs in several aspects. One of the important issues in this regard is the interference to other wireless networks or Primary Users (PUs). Protecting the right of Primary users is the major concern of the CR -WSN. Therefore, the miss detection probability of PUs should be minimized in order to minimize the interference with the PUs. The false alarm probability should also be minimized as large false alarm rates will also cause spectrum to be under-utilized. High false alarm and miss detection probabilities in the CR-WSNs in environment monitoring scenarios should be deeply studied as this can dramatically degrade the performance of wireless sensor networks for surveillance purposes.

6.2.3 Joint Communication and Sensing in One Technology for WSNs

Integrated sensing and radio communication systems have emerged in sake of system miniaturization and transceiver unification. With the current technological advancements the radio frequency front-end architectures in sensing networks and radio communications become more similar. Orthogonal Frequency Division Multiplexing (OFDM) as a capable technology already and successfully used in wireless communications (e.g., in IEEE802.11a,g,n,p standards) can be used in wireless sensor networks.

Among the state-of-the art transmission schemes the seminal concept of joint ranging (location) and communication using OFDM technology [3] is important. In this context OFDM can find applications in future WSNs where context-awareness (location information) is important and, in addition to sensing, data exchange among nodes or between nodes and fusion center is essential.

In addition to reconfigurability and adaptation capability of OFDM, by using this technology besides achieving both functionalities simultaneously, the bandwidth will be efficiently used. Furthermore, by using the OFDM technique, in high mobility scenarios, the Doppler effect can be detected in real time. This technology is for example favourable in a UAV sensing network monitoring large areas where the communication channel between cluster head node and the fusion center can be highly mobile.

6.2.4 Security of WSNs

Security is an extremely crucial issue in various applications of WSNs. This is because the sensor nodes are prone to various kinds of malicious attacks. Therefore, security mechanisms will be required to safeguard transmission of sensed data. Security of WSNs is remarkably different from traditional secure communication systems. This is because sensor nodes of a WSN deployed in the region of interest interact with the environment and broadcast data. Therefore, security mechanism of WSNs seems different from traditional security of communication systems (transmitting signal from a transmitter to the receiver). Major challenges for security of WSNs is the physical limitations of sensor nodes in terms of size, battery power and memory which make the complete implementation of security mechanisms (encryption and decryption) a hard task. Furthermore, as mentioned above, in WSNs nodes broadcast the sensed data in a network. Therefore, there will be more chance for collision of data. Moreover, in a hostile environment the probability of trapping the sensed data is obviously higher. For the security of WSNs sensor nodes and base stations should authenticate the received data to verify whether they are sent from a trusted node. In the protocol layer of WSNs secure management mechanisms should be considered as encryption and routing information require high level of security management. For a full security at each sensor component of a WSN a security mechanism must be implemented which with regard to the limitations of sensor nodes is by far a real challenge.

6.2.5 Cooperative Aspect of WSNs

The cooperative scheme in WSNs can effectively increase the quality of service of the network when compared to non-cooperative methods. Many challenges in applying Game Theory to cooperative wireless sensor networks are envisioned. Among them theoretical methods to design cooperative WSNs, study of cooperation schemes, protocol design, and deployment of cooperative networking over existing infrastructure are just a few to name. In the next section, we focus on the game-theoretic aspect of cooperative WSNs, especially on dynamic node's coalition formation algorithms, and see how this can be formulated as a Game Theory problem.

6.3 A Game-Theoretic Framework for WSN

As discussed above, the need for self-organizing, sometimes decentralized and autonomous networks, make suitable game-theoretical tools of paramount interest. These tools can be used to analyse the behaviour and interaction of nodes in WSN. As already mentioned, in a WSN a few nodes may cooperate to increase the performance of the system. In this paper, we use game theory, to study the nodes' cooperation as a process of arriving at coalitions within WSN. Such a coalition of cooperating nodes is also called a cluster. These coalitions have to be flexible, have reconfigurable structures that can adapt to the circumstances, and also should be able to cooperate with other sensor networks. In studies on the process of coalition formation an important issue is how coalitions form, or by which rules they proceed before some kind of stability is reached. In the more classic approach of game theory, it is assumed that (for some reason) the grand coalition is formed, consisting of all players (nodes). Subsequently, the gain is divided between all players (nodes) is such a way that players do not have an incentive to leave or defect form this grand coalition and form other coalitions instead, see [4, 5]. In that approach the issue of how coalitions form is not taken into consideration. But in forming coalitions for optimizing some task, like detecting or tracing an intruder using a WSN, generally this grand coalitions is inefficient due to the costs of information exchange between the nodes. Instead some partitioning of nodes will emerge, where nodes work together in coalitions to perform tasks. In addition, in dynamic scenarios, the partitioning of the total set of nodes in a WSN may change over time. Therefore, in applying game theory to WSN, one has to describe how coalitions form. In the literature one may find models in which the formation of coalitions may be constrained by an underlying network, connecting the various nodes. It may even be the case that links are formed or deleted, like in social networks, where nodes try to obtain unique positions. In social networks analysis for example, the research question is: what kind of networks do emerge, what is its topology, see [10] or, for an overview [11]. In the discussion that follows here, we do not take these kinds of constraints into consideration, except for the case that some coalitions may need the presence of a node that transmits information to a base station.

6.3.1 Aspect of Game Theory Relevant for Cooperative WSN

Coalition formation entails finding a coalitional structure which maximizes some technical welfare or utility. This technical welfare can be defined in various ways, depending on the application at hand. Finding the structure that has the maximum welfare requires iterating over all partitions of nodes, each time determining its technical welfare. But, as for 10 nodes there already exist more than 115000 partitions, this centralized approach is computationally complex and impractical. Therefore, some specific coalition formation algorithms are used, consisting of simple rules for forming or breaking coalitions. After several iterations of these rules, the stability of the obtained partition of the set of all nodes is assessed. In our presentation, we closely follow the approach of [6, 7] and [12].

6.3.2 Technical Welfare

Let N be the set of nodes in a WSN, and let $| N |$ be its size. A coalition (a cluster of nodes) is a non-empty subset of N. We assume that some value can be assigned to such a coalition. For a coalition S, this value is noted by $v(S)$. This value models the trade-off between the advantage of cooperation (in terms of better performance) and the costs of cooperation (in terms of bandwidth, transmitting power, or exchanging information). The value may depend on the topology within such a coalition (for example the presence of an information fusion centre), node's location within the area of interest, and on characteristics of the individual nodes.

A collection of the (grand) coalition N is any family $S = \{S_1, S_2, \ldots, S_t\}$ of mutually disjoint coalitions of N; its size is t. If, in addition, the union of these disjoint coalitions is N, it is called a partition of N. In forming flexible structures that can adapt to the circumstances, the set of nodes is partitioned into mutually disjoint coalitions. These mutually disjoint coalitions may be interpreted as alternatives structures that can perform some task that the WSN is constructed for. In the process of finding an optimal coalitional structure, a particular partition of nodes is compared to alternative partitions by means of a technical welfare function. Given a partition $\{S_1, S_2, \ldots, S_t\}$, this technical welfare is denoted by $tw(S_1, S_2, \ldots, S_t)$, which is a function of the values $(v(S_1), v(S_2), \ldots, v(S_t))$. This can also be written as $tw(v(S_1), v(S_2), \ldots, v(S_t))$. Technical welfare of two partitions can be compared in various ways. We give a few examples. Let $S = \{S_1, S_2, \ldots, S_t\}$, and $T = \{T_1, T_2, \ldots, T_k\}$ be two partitions of the set of nodes.

- *Utilitarian comparison*
  S is preferred to T if $\sum_{i=1}^t v(S_i) > \sum_{j=1}^k v(T_j)$, so $tw(x_1, \ldots, x_t) = \sum x_i$
- *Nash comparison*
  S is preferred to T if $\prod_{i=1}^t v(S_i) > \prod_{j=1}^k v(T_j)$, so $tw(x_1, \ldots, x_t) = \prod x_i$

- *Maximum comparison*

  S is preferred to T if $\max_i v(S_i) > \max_j v(T_j)$, so $tw(x_1, \ldots, x_t) = \max_i x_i$

Which comparison function is used depends on the application in mind. Technical welfare combines coalitional values with the value generated by the WSN as a whole. (As an aside: if one also distributes the value $v(S)$ between the nodes of S, transforming the vector $v(S_1), v(S_2), \ldots, v(S_t)$ into a vector $(s_1, s_2, \ldots, s_{|N|})$ with $|N|$ entries, one may also use the *Pareto comparison*: S is preferred to Q if $(s_1, s_2, \ldots, s_{|N|})$ is preferred to $(q_1, q_2, \ldots, q_{|N|})$, meaning that $s_i = q_i$ and there exists an index j such that $s_j > q_j$.)

6.3.3 Transitions between Partitions

Now we come to the issue of how coalitions of WSN nodes form, or evolve over time. A transition from one partition to another one is feasible, or is likely to happen, if the technical welfare increases. To describe the possible alternative partitions, we need some simple yet rich enough rules by which one may proceed towards optimal partitions. Again, there are several choices for these rules that modify an existing partition. We give three examples of such rules that may be used to transition to a next partition.

- Merging coalitions

  $\{Z_1, Z_2, \ldots, Z_k\} \cup S$ becomes $\bigcup_{j=1}^{k} Z_j \cup S$ if $tw\left(\bigcup_{j=1}^{k} Z_j\right) > tw(Z_1, Z_2, \ldots, Z_k)$

- Splitting a coalition

  $\bigcup_{j=1}^{k} Z_j \cup S$ becomes $\{Z_1, Z_2, \ldots, Z_k\} \cup S$ if $tw(Z_1, Z_2, \ldots, Z_k) > tw\left(\bigcup_{j=1}^{k} Z_j\right)$

- Exchanging

  $\{Z_1, Z_2\} \cup S$ becomes $\{(Z_1 \backslash U_1) \cup U_2, (Z_2 \backslash U_2) \cup U_1\} \cup S$ where $U_1 \subset Z_1$ and $U_2 \subset Z_2$, if $tw(Z_1, Z_2) < tw((Z_1 \backslash U_1) \cup U_2, (Z_2 \backslash U_2) \cup U_1)$

Clearly, this definitions assume that the technical welfare of a partition can be built from technical welfares of parts of the WSN. For example, if $tw\left(\bigcup_{j=1}^{k} Z_j\right) > tw(Z_1, Z_2, \ldots, Z_k)$, then $tw\left(\bigcup_{j=1}^{k} Z_j, S\right) > tw(Z_1, Z_2, \ldots, Z_k, S)$ for any S. See [5] for introducing the rules above and discussing this issue of monotonicity. The rules may, in some sense, be compared to defining neighbourhoods of a given point in a solution space if one uses heuristics for optimization, like simulated annealing.

To illustrate the use of the rules above, and investigate the sufficiency of these rules for reaching optimal partitions, we consider the following example from [5]. Let $N = \{1,2,3,4\}$, and let v be defined by:

$$v(S) = \begin{cases} 3 & if\ S = \{1,2\} \\ |S| & otherwise \end{cases}$$

If we take the utilitarian comparison, the partition with the highest *tw*-value is $\{\{1,2\}, \{3,4\}\}$, having a value of 5. Now start with any partition. Using the merging and splitting rule, we arrive at either $\{\{1,2\}, \{3,4\}\}$ or at other partitions with two coalitions, where each coalition has two elements, like $\{\{1,3\}, \{2,4\}\}$. These partitions all have a value of 4. Note that if we only apply the merging and splitting rules, it is impossible to transit from $\{\{1,3\}, \{2,4\}\}$ to the partition with value 5. For that to happen, we in addition need the exchange rule. Unfortunately, one may construct examples, where even the addition of this rule does not make it possible to reach the partition with the highest *tw*-value.

6.3.4 Testing Stability; D-Stable Partitions

In the classical game-theoretic approach, one assumes that the grand coalition of nodes will be formed. Then, a proposed solution for dividing the gains of cooperation has to be tested with respect to stability. A well-known concept then is the core of the game, consisting of payoffs for which no coalition of player can obtain a better payoff by leaving the grand coalition.

So far, we described transitions from one partition to another one by means of rules, like splitting and merging. To test the stability of a particular partition that may have emerged after iterations of the merging and splitting rules, we use the notion of defecting functions D. This defection function can be compared with testing whether or not the payoff structure within the grand coalition belongs to the core. Given any partition S of the nodes, the set $D(S)$ consists of partitions that serve as alternative to the existing partition. An example is the set D_{hp}. Here $D_{hp}(S)$ consists of partitions that can be reached from S in just one iteration of the merge or split rules. Repeated application of merge/split, starting at any partition, always terminates (due to the finite number of partitions) at what is then called a D_{hp}-stable partition. As illustrated in the example above, the outcome is not always unique, nor does splitting and merging guarantee maximal utility.

Another example is the set $D_c(S)$ consisting of all possible collections of coalitions, not necessarily a partition. Unfortunately, D_c-stable partitions may not exist. But, if a D_c stable partition exists, it is the unique outcome of applying merging and splitting to any partition. Given a partition S, we may also consider transitions due to a defecting collection T, where T is not necessarily a partition. Let $T = \{T_1, T_2, \ldots, T_k\}$, and let $\hat{T} = \bigcup T_i$. The collection may be evaluated in its defected form as $tw(T_1, T_2, \ldots, T_k)$. But in the frame of $S = \{S_1, S_2, \ldots, S_t\}$, the current partition, it may be evaluated as $tw(S_1 \cap \hat{T}, \ S_2 \cap \hat{T}, \ \ldots S_t \cap \hat{T})$. Note that if $\hat{T} = N$, then these two evaluations are equal. A transition is possible if $tw(T_1, T_2, \ldots, T_k) > tw(S_1 \cap \hat{T}, S_2 \cap \hat{T}, \ldots, S_t \cap \hat{T})$. As generally speaking no partition S is immune for transitions including coalitions, we restrict ourselves to transitions between partitions.

6.3.5 An Illustrative Example (WSN for Intruder Detection Scenario)

In this application we focus on the monitoring and surveillance of a region and detection of an intruder in this region using a cooperative WSN. In this application there are several choices for assigning value to a coalition, as well as assigning the technical welfare to a collection or partition. Giving a task that the WSN has to perform, like detecting an intruder, the value v models the trade-off between the advantage of cooperation (in terms of better detection performance) and the costs of cooperation (in terms of bandwidth and power). The technical welfare function is tuned to resemble the quality of the partition in performing a task. In this example, we consider the task of detecting an intruder, see Figure 6.2.

The gain or quality of cooperation $Q(M)$ of a coalition M depends on how the coalition of nodes is able to distinguish the intruder's signal A from noise. In our (distributed) detection case, this gain is not a simple function. The ith node's Maximum Likelihood detection performance $P_{d,i}$ is:

$$P_{d,i} = Q\left(\frac{Th - A}{\sigma}\right)$$

where Th is threshold level of detection of the signal and σ^2 is the variance of noise. The ith node's Probability of False alarm $P_{fa,i}$ can be written as:

$$P_{fa,i} = Q\left(\frac{Th}{2\sigma}\right)$$

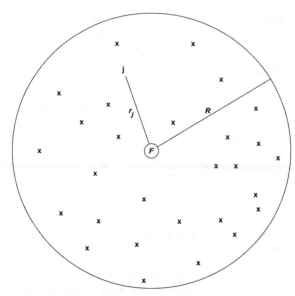

Figure 6.2   WSN configuration for intruder detection. R is the coverage of WSN; F is the fusion node; the x's are typical nodes of a WSN; r_j is the distance of node j to the fusion node.

Where Q(.) is:

$$Q(x) = \frac{1}{\sigma\sqrt{2\pi}} \int_x^\infty e^{-\frac{z^2}{2\sigma^2}} dz$$

For the distributed detection the total Probability of Detection P_D, and P_{FA} are [14]

$$P_D = \sum_{\Lambda(r)>V_*} Prob(\Lambda\,(\mathbf{r})\,|H_1)$$

$$P_{FA} = \sum_{\Lambda(r)>V_*} Prob(\Lambda\,(\mathbf{r})\,|H_0)$$

where V_* is the threshold, $\Lambda\,(r)$ is the Liklihood ratio and for a network of M independent sensor signals $r = [r_1 r_2, \ldots, r_M]$ is:

$$\Lambda\,(\mathbf{r}) = \prod_{i=1}^{M} \frac{Prob\{r_i|\,H_1\}}{Prob\{r_i|\,H_0\}}$$

H_1 and H_0 are respectively the hypothesis that signal is present and absent. The Likelihood ratio $\Lambda\left(r\right)$ should be compared with a Threshold vector (V_*) which has M components. It can be shown that:

$$Prob\left(log\ \Lambda(r_i)|\ H_1\right) = (1 - P_{d,i})\ \delta\ \left(log\Lambda(r_i) - log\frac{1 - P_{d,i}}{1 - P_{fa,i}}\right)$$

$$+ P_{d,i}\delta\left(log\Lambda(r_i) - log\frac{P_{d,i}}{P_{fa,i}}\right)$$

$$Prob\left(log\ \Lambda(r_i)|\ H_0\right) = (1 - P_{fa,i})\ \delta\ \left(log\Lambda(r_i) - log\frac{1 - P_{d,i}}{1 - P_{fa,i}}\right)$$

$$+ P_{fa,i}\delta\left(log\Lambda(r_i) - log\frac{P_{d,i}}{P_{fa,i}}\right)$$

where $\delta\left(.\right)$ is the Dirac Delta function. The above couple of equations explain clearly the performance quality of $Q(M)$ when M sensor nodes cooperate in the detection of intruder. As shown above the performance depends on the probabilities of detection, false alarms and thresholds.

Now we focus on the cost of cooperation $C(M)$. For each node is given p_i, the power of the node. The parameter r_i is the distance of node i to the fusion center (see Figure 6.2), R is the coverage of WSN, while σ^2 is the noise variance. Then we define the cost of cooperation within the coalition M by

$$C\left(M\right) = \sum_{i=1}^{M}\frac{r_i^2}{R^2} + \log\left(1 + \frac{\sum_{i=1}^{M}\frac{p_i}{r_i^2}}{\sigma^2}\right).$$

As a value for the coalition M we may take

$$v(M) = \alpha Q(M) + (1 - \alpha)\,C\left(M\right), \quad \text{where} \quad 0 < \alpha < 1,$$

thereby balancing costs and performance or quality. This approach may be used in simulations to investigate the flexible adaptation of coalitions to tasks and changing circumstances. This dynamic coalition formation process, with distributed nodes and a fusion center, is based on iterated application of the merging and splitting rules.

6.4 Summary

Game theory is used in studying cooperative wireless networks for two reasons [13]. The first is to understand incentives and strategies of nodes for cooperation. The second one is to devise distributed algorithms that can find optimal partitions of the set of nodes. Our distributed approach using merging and splitting can be suboptimal, but runs in polynomial time, contrary to a more centralized approach.

Following the formulation of the problem, in the future research, we will start simulations, using several scenarios, with nodes having certain distinct characteristics. This will be used to validate the performance of the game-theoretic approach in various WSN application scenarios. There we will also elaborate on the gain of cooperation, which will require more sophisticated probabilistic arguments.

References

[1] H. Nikookar, Wireless Radio Senor Networks: Looking Back, Moving Forward, Seminar on Human Bond Communications and Beyond 2050, Aalborg University, Denmark, June 2015.

[2] Gupta and Kumar, Wireless Sensor Networks: A Review, International Journal of Sensors, Wireless Communications and Control, 2013, Vol. 3, No. 1.

[3] P.v. Genderen and H. Nikookar", Radar Network Communication", 6th International Conference on Communications, June 2006, Bucharest, Romania, pp. 313–316.

[4] L. Evers, AI. Barros, H. Monsuur (2013). *The Cooperative Ballistic Missile Defence Game.* In: Das SK, Nita-Rotaru C, and Kantarcioglu M (eds). *GameSec 2013*, LNCS 8252, 85–98.

[5] T. J. Grant, R. H. P. Janssen, H. Monsuur (eds). (2014). *Network Topology and Military C2 Systems: Design, Operation and Evolution.* IGI Global Publishers.

[6] K. Apt and A. Witzel, "A generic approach to coalition formation", in Proc. of the Int. Workshop on Computational Social Choice (COMSOC), Amsterdam, the Netherlands, Dec. 2006.

[7] K. Apt and T. Radzik, Stable partitions in coalitional games. arXiv:cs/0605132v1 [cs.GT], May 2006.

[8] P. Bejar, P. Belanovic and S. Zazo. Cooperative localization in wireless sensor networks using coalitional game theory. Proceedings 18th European Signal Processing Conference, August 2010.

[9] S. Kim. Game Theory applications in Network Design.

[10] H. Monsuur (2007). *Stable and Emergent Network Topologies: A Structural Approach.* European Journal of Operational Research, 183(1), 432–441.

[11] M. O. Jackson (2008). Social and Economic Networks. Princeton University Press.

[12] W. Saad, Z. Han, M. Debbah, A. Hjørungnes, T. Basar (2009). Coalitional Game Theory for Communication Networks: A tutorial. IEEE Signal Processing Magazine, 26(5): 77–97.

[13] A. B. MacKenzie, L. A. daSilva (2006) Game theory for wireless engineers, in: Synthesis Lectures on Communications, Tranter, W, Ed.

[14] S. Thomopoulos, R. Wisanathan and D.C. Bougoulias, "Optimal Fusion in Multiple Sensor Systems," IEEE Trans. Aerospace and Electronic Systems, vol. AES23, no. 5, Sept. 1987, pp. 644–653.

Biographies

Homayoun Nikookar received his Ph.D. in Electrical Engineering from Delft University of Technology in 1995. He is an Associate Professor at the Faculty of Military Sciences of the Netherlands Defence Academy. In the past he has led the Radio Advanced Technologies and Systems (RATS) research program, and supervised a team of researchers carrying out cutting-edge research in the field of advanced radio transmission. His areas of research include Wireless radio channel modeling, Ultra Wideband (UWB) Technology, MIMO, Multicarrier-OFDM transmission, Wavelet-based OFDM en Cognitive Radio Systems and Networks.

He has received several paper awards at international conferences and symposiums. Dr Nikookar has published 150 papers in the peer reviewed international technical journals and conferences, 15 book chapters and is author of two books: *Introduction to Ultra Wideband for Wireless Communications*, Springer, 2009 and *Wavelet Radio: Adaptive and Reconfigurable Wireless Systems based on Wavelets*, Cambridge University Press, 2013.

Herman Monsuur is a Professor at the Faculty of Military Sciences of the Netherlands Defence Academy. He studied (pure) Mathematics at the University of Groningen, the Netherlands and in 1994 received his PhD from Tilburg University. He is also head of the Expertise Centre Military Operations Research where he tries to solve operational and logistic problems using scientific OR methods (network theory, game theory, decision theory, optimization, and modeling and simulation), with the goal of optimal operational deployment and readiness.

7

EASY-PV: RPAS Professional Business Application

Bilal Muhammad and Ramjee Prasad

CGC, Department of Business Development and Technology,
Aarhus University, Denmark

Abstract

Remotely Piloted Aircraft System (RPAS) or Unmanned Aerial Vehicle (UAV) coupled with sensor and global positioning technology is disrupting existing businesses models in numerous ways. Infrastructure inspection and maintenance using RPAS is the largest market segment of RPAS technology. This chapter presents EASY-PV, a practical solution for the automatic maintenance of utility-scale photovoltaic power plants by employing RPAS.

Keywords: Remotely piloted aircraft, unmanned vehicle, search and rescue, infrastructure, EASY-PV.

7.1 Introduction

RPAS has made significant ground in a multitude of businesses that include, but not limited to, Agriculture, Energy, Infrastructure inspection and Search and Rescue (SAR). Due to the customization flexibility of payload, RPAS offers enhanced work efficiency and productivity, reduced operational costs, refined service and customer relations, on a large scale to businesses globally. This paved the way for RPAS commercial market to reach approximately 127 billion U.S. dollars by 2025 according to Business Insider. The rapidly growing investment in commercial RPAS generated a greater demand for research and innovation in respective UAV business application areas.

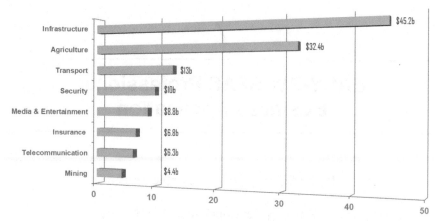

Figure 7.1   RPAS Product and/or Service Expected Market Value.

Figure 7.1 shows the expected market value of each sub-segment of RPAS product and/or service. The infrastructure inspection market sub-segment holds the largest share followed by agriculture and transportation industry and so on.

On the face of green-energy initiatives, investment in photovoltaic power plant infrastructure is ever growing. This allowed the PV industry to expand immensely ranging from smaller user segments to larger PV plants that generate 150 Megawatt or more.

PV Industry is reaching the "age of maturity" with respect to existing PV plants, therefore further attention is being given towards maintenance technologies and processes in order to enhance energy production. Though significant improvements have been made in enhancing the performance of PV systems, nevertheless, PV panels experience aging over a period that result in faults, more common among them are the hotspots, which are mostly not visible to naked eye.

Thermal anomalies or hotspots on PV plant modules as shown in Figure 7.2, decrease the efficiency of an entire modules string during the conversion from solar energy to electricity; therefore, the requirement of identification and replacement of faulty modules is strongly requested by companies operating in the field of maintenance due diligence of photovoltaic systems and by PV plants owners. Indeed, several factors are affecting the severity of PV modules degradation:

Figure 7.2 Sample of PV modules defects—Visual and Thermal View.

- Production defects
  - Manufacturing materials quality
  - Manufacturing process
  - Quality of assembly and packaging of the cells into the module
- Transport and installation damages
- Environmental conditions
  - Normal conditions (temperature, humidity, pollution)
  - Extreme events (wind, hail, big snow)

Thermo-diagnosis or Infrared (IR) Imaging using IR camera is one of the most effective methods for detecting hotspots as shown in Figure 7.2. The inspection of individual PV panels however can be time consuming and expensive depending on the size of PV plant. The conventional methods of thermos-diagnosing using handheld cameras are operationally complicated, manual, and time-consuming, particularly when the PV plants are located on rooftop.

EGNSS High Accuracy System for the Maintenance of Photovoltaic Plants (EASY-PV) [1, 2] is an automatic solution for the inspection of utility-scale photovoltaic plants in a cost and time effective manner without any human intervention. EASY-PV employs RPAS with a customized payload that performs a flying mission over the photovoltaic power plant. The RPAS collects thermal images tagged with high accurate positioning information processed with a specialized computer vision algorithm would allow to identify, detect, and precisely locate the defective solar panel.

This chapter provides an overview of the EASY-PV system architecture, subsystem design, and potential users and customers.

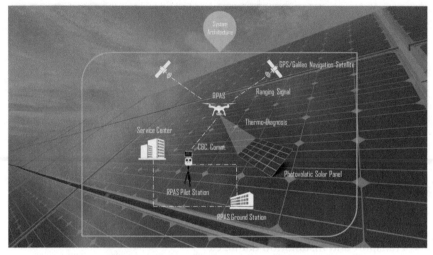

Figure 7.3   EASY-PV System Architecture.

7.2 EGNSS High Accuracy System for Improving Photovoltaic Plant Maintenance

7.2.1 EASY-PV System Architecture

EASY-PV system architecture is presented in Figure 7.3. As depicted, the proposed solution employs RPAS, which is equipped with a thermal camera, GNSS RTK receiver, and visual sensor. The RPAS shall perform a mission flying over a photovoltaic field to collect optical and thermal images of solar panels in a PV plant. The data collected data by RPAS is processed by means of a computer vision algorithm, and the captured images are tagged by means of centimeter-level position fix provided by RTK GNSS receiver. The centimeter-level positioning provided by GNSS enables the automation of the entire process allowing to correctly geo-reference the defective panel inspected by the on-board thermal camera. Finally, this information shall be provided to the remote service center in charge of thermal anomalies identification and management.

7.2.2 EASY-PV System Design

The proposed system is composed of three subsystems:

- Service center
- RPAS
- RPAS Ground Station

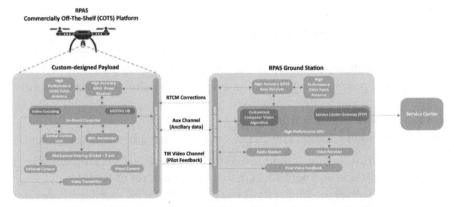

Figure 7.4   RPAS and RPAS Ground Station Subsystem Design.

The block diagram of EASY-PV subsystem design is presented in Figure 7.4. Each of the subsystems is briefly discussed in the following subsections.

7.2.2.1 Service Center
Service center is devoted to manage all aspects related to production, delivery, archiving and cataloguing of data acquired and processed. It also performs RPAS flight mission planning and cooperation between the various entities involved.

7.2.2.2 RPAS
RPAS is composed of COTS (Commercially Of the Shelf) or a self-built drone platform, Communication and Control (C&C) unit, and custom designed payload.

The PRAS custom designed payload consists of the following subsystem components.

A. GNSS Receiver equipped with Tallysman TW2710 antenna. The receiver is ublox M8P [3], which supports dual GNSS constellation and RTK operation. The receiver offers centimeter-level positioning using a local base-rover approach with the functionality of transmitting/receiving RTCM corrections from/to another ublox M8P receiver over a UHF link.

B. Thermal Camera FLIR VUE PRO for thermal imaging of photovoltaic panels.

C. 2-axis mechanical gimbal with its dedicated electronic control unit that is designed for handling thermal and visual sensors.

D. On-Board Computer (OBC) is a general-purpose processor that interfaces the thermal sensor, visual sensors and GNSS receiver, performs data and position logging and on-board components synchronization.

7.2.2.3 RPAS Ground Station

The RPAS Ground Station consists of the following subsystem components:

A. RPAS Pilot Station provides real-time thermal and visual feedback of RPAS mission and performs command and control of RPAS.

B. GNSS receiver module at RGS is a ublox M8P RTK receiver. It operates as a local base station providing RTK correction in RTCM (v.3.x) format over a UHF radio link to M8P rover integrated in RPAS payload.

C. Radio Modem is used for receiving real-time video and data reception.

D. Central Processing Unit (CPU) with dedicated Graphical Processing Unit (GPU) to perform computer vision operations [4] on the RPAS collected data. As a first choice, Intel Joule 570x development board is selected. The CPU also provides a gateway interface to Service Center.

7.2.3 EASY-PV Users & Customers

EASY-PV solution targets the maintainers and/or owners of utility-scale photovoltaic power plants as final users, which are encouraged to use customized RPAS payload design together with RPAS ground segment and a dedicated service center to enhance PV plant power production in a cost and time effective manner with no direct human intervention.

Two unique business models can be envisaged:

- The EASY-PV service model
- The EASY-PV product model.

Within the *service model*, the goal is to offer a complete service to final customers – especially domestic plant owners and small maintainers not so familiar with RPAS technology – that will basically fully outsource to EASY-PV the analysis of faulty panels.

The *product model* is primarily intended for big players (maintainers) that usually manage a relevant number of plants and are willing to invest in capital expenditures in order to operate independently from an external service. Such model refers indeed to the commercialisation of the payload to be mounted on the RPAS and the RPAS Ground Station but also, and most importantly, a license scheme for the Maintenance and Control (M&C) platform capable of programming the RPAS flight route, providing the post-processing of the images taken from the RPAS and generating the report on faulty panels. This license scheme has been determined by EASY-PV consortium equal to €200 per each MW, up to €30.000 per year in case of a total volume larger than 150 MW. Such model will likely require a strong commercial effort for offering the solution and providing the customer support whether directly or indirectly through agreements with resellers.

7.3 Conclusion

EASY-PV solution presented in this chapter automates the maintenance of PV power plant thus enhancing PV plant power production in a cost-effective manner by minimizing human intervention. As a first step towards a complete market-ready solution aligned with PV maintenance user needs, EASY-PV provides end-to-end approach to identify, detect and precisely locate faulty PV panel in a large-scale photovoltaic plant. The solution can be introduced in the market either as a service model or a product model depending on the user needs and requirements.

References

[1] Bilal Muhammad, Ramjee Prasad, Marco Nisi, Alberto Mennella, Graziano Gagliarde, Ernestina Cianca, Davide Marenchino, Marioluca Bernardi, Addabbo Pia, Ullo Silvia "Automating the Maintenance of Photovoltaic Power Plants" Proceedings of IEEE Global Wireless Summit, 2017

[2] Marco Nisi, Fabio Menichetti, Bilal Muhammad, Ramjee Prasad, Ernestina Cianca, Alberto Mennella, Graziano Gagliarde, Davide Marenchino "EGNSS High Accuracy System Improving Photovoltaic Plant Maintenance using RPAS integrated with Low-cost RTK Receiver" Proceedings of Global Wireless Summit, 2016.

[3] Mongrdien, C., Doyen, J.-P., Strom, M. and Ammann, D. (2016). "Centimeter-Level Positioning for UAVs and Other Mass-Market Applications," in Proceedings of the 29th International Technical Meeting of the Satellite Division of the Institute of Navigation (ION GNSS+), Portland, Oregon.

[4] GaryBradski and Adrian Kaebler, "Learning OpenCV: Computer vision with the opencv library," OReilly, 2008.

Biography

Bilal Muhammad holds a Ph.D in Telecommunication Engineering from the University of Rome Tor Vergata, Italy. He is working as a postdoc at the Department of Business Development and Technology, Aarhus University. His research focus is on the use of GNSS in aviation and mass-market user applications. Currently, he is involved in EU H2020 projects focusing on the application of Remotely-Piloted Aircraft System in a multitude of businesses, which include, Inspection and Monitoring, Search and Rescue, and Precision Agriculture.

8

The Situation of Network Neutrality in Service Innovation Era

Yapeng Wang and Ramjee Prasad

CGC, Aarhus University, Denmark

Abstract

In 2015, Federal Communication Committee (FCC) and European Commission enacted respective rules in relation to open internet, the network neutrality (as commonly known) or Net Neutrality (as in Europe) was gained more attentions and the topic was discussed ardently. On December 14, 2017, the FCC Committee voted 3:2 to repeal "Internet Open Regulations". This chapter reviews the development of NN debate process, and the opinions from different sides, including the network providers, the service providers, other relevant companies, governments and researchers. This chapter also introduces a telecommunication convergence concept that is CONASENSE(Communication, Navigation, Sensing and Services). It aims to formulate a vision on solving societal problems with new telecom technique to improve human welfare benefit. This chapter focuses on the service of CONASENSE and summarized the current situation of NN in service innovation era.

Keywords: Network Neutrality, CONASENSE, Innovation, Long term benefit.

8.1 Introduction

The Information and Communication Technology (ICT) is playing an absolutely important role in our life and with more and more innovative internet services and applications emerge, from Ecommerce, e-health to a real-time

Towards Future Technologies for Business Ecosystem Innovation, 117–136.

telephone meeting and a live video streaming, which are improving human's Quality of Life and benefiting the whole society. We have entered a service innovation era, in which every part of the industry chain is making contribution to the telecom industry. Internet Services Providers (ISPs) are continuing to upgrade the network, the Content Providers (CPs) are making several contributions to keep the telecom industry prosperous. How these innovative service run over the telecommunication network is governed by not only technology, but also by the rules as proved by government or authorities in some region. NN, as a regulation, aims that every end user have the equal right to access the internet and use the legal internet content and applications. Since On December 14, 2017, the Federal Communications Commission (FCC) abolished Obama's neutral network policy and re-empowered telecom operators with control over broadband Internet access. This neutral network policy was repealed and it can also be regarded as the official opening of the Light-touch Regulation era in the United States. CONASENSE, as will be discussed later in this chapter, is a telecom convergence concept that will be run above ICT platform. This chapter will focus on the service in CONASENSE, and analyze the abolishment of NN rules impact on this service.

8.1.1 NN

Network Neutrality rules aims to provide an open internet [2] to the end users. The open internet is defined by FCC and refers to *"uninhibited access to legal online content without broadband Internet access providers being allowed to block, impair, or establish fast/slow lanes to lawful content"* [1].

This means that a legitimate content, whether it is an application or data, must reach users without the intermediate communication system controlling its flow. Internet Content Providers (ICPs) and Internet Service Providers (ISPs) cannot block, throttle or create the special facilities for a content or application. This gives a kind of liberty to the end users to enjoy the variety of information without bothering about how ICT is dealing with that information. Therefore, users may demand a better network services to enjoy the lawful contents. Section 8.2 will introduce NN in detail.

8.1.2 Innovative Services

In this chapter, the innovative service refers to over the top service (OTT) which implies communications carried over the physical network infrastructure using an IP protocol to reach services available on the internet [3].

An OTT application is any app or service that provides a product over the Internet and bypasses traditional distribution. Services that come over the top are most typically related to media and communication and are generally, if not always, cheaper than the traditional method of delivery. Section 8.4 will discuss the change of NN's impact on it.

8.1.3 CONASENSE [4]

CONASENSE foundation was established as a brain tank in November 2012. CONASENSE refers to the Communication, Navigation, Sensing and Services. Its main aim is to define and steer processes directed towards actions on investigations, developments and demonstrations of novel CONASENSE services, as an integration of communications, navigation and sensing technology, with high potential and importance for society. CONASENSE has a large research scope and it has a huge capacity to develop various application and service provision to users. Section 8.4 will give a detail introduce to this concept.

This chapter points out that both sides (both sides refer to the opponent and proponent of NN in this chapter) of the NN debate agree on the need to preserve the Internet as a space that is open to innovation, and the freedom of users to access the content and services. Meanwhile this chapter also analyses

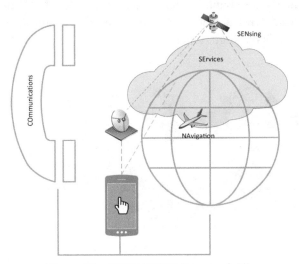

Figure 8.1   CONASENSE Framework [4].

how NN impact innovation such as CONASENSE, and meet the regulation goal. This chapter makes analysis mainly from technique perspective, and the social and economic aspects are also discussed.

Apart from Section 8.1, the rest of the chapter is organized as follows: Section 8.2 reviews the evolution and current discussions of NN, Section 8.3 briefly introduces CONASENSE, Section 8.4 qualitatively analyses the impact of the abolishment NN on Services and finally Section 8.5 concludes the chapter.

8.2 The Network Neutrality

In the world, more and more regulators believe that the CPs are the motivation of innovation, economy and investment, but with the rapid development of the Internet as an ubiquitously available platform and resource, the network infrastructure owners are regarded as to the broadband providers are regarded as to *"have both the incentive and the ability to act as gatekeepers standing between edge providers and consumers. As gatekeepers, they can block access; target competitors, extract unfair tolls"* [1], but the open internet is regarded to the guarantee to the innovation, economy and investment. This is one of the vital reasons why more and more counties government to enact the NN rules. Presently, more than 10 countries enacted relevant rules [5]. Figure 8.2 shows the countries that either have enacted NN or are in their discussion phase. On December 14, 2017, the FCC Committee voted 3:2 to repeal "Internet Open Regulations."

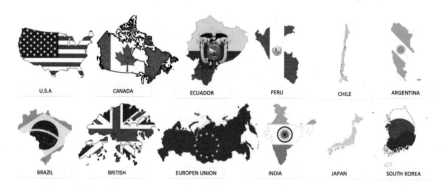

Figure 8.2   The Countries that either have enacted or are in their discussions.

8.2.1 The Concept NN

The concept of NN commonly indicates that Internet services providers make or keep the Internet open and ensure all the users have same right to access to the network and use the content and services without any discrimination [1].

8.2.2 The Principles of NN

Currently, Network Neutrality is a global debate [5]. FCC released its updated Open Internet Order in 2015, to enact strong, sustainable rules to protect the Open Internet. The order includes 3 bright-line rules as [1]:

- No Blocking (NB), to prohibit the network providers to block the legal applications, contents and devises.
- No Throttling (NT), to prohibit the network providers to degrade the traffic of legal applications and contents.
- No Paid Prioritization (NPP), to prohibit the network providers to provide and charge the differentiated service for the applications and contents.

"No unreasonable interference or disadvantage to consumers or edge providers" [1], and, enhanced transparency between users and ISPs. "*As with the 2010 rules, this Order contains an exception for reasonable network management, which applies to all but the paid prioritization rule*" [1] is also emphasized in FCC rules.

Nonetheless, in October 2015, the European Parliament approved the first EU-Wide rules on NN that enshrined its principle into EU law [6] that says: "*No blocking or throttling of online content, applications and services. Accordingly, every European must be able to have an access to the internet and, all contents and services, via a high-quality service that is provided by an ISP such that all traffic must be treated equally. NN rules seems more inclined to the end users and, equal treatment allows reasonable day-to-day traffic management according to justified technical requirements which must be independent of, the origin or destination of the traffic, and, of any commercial considerations.*"

8.2.3 The State of the Art of NN

On December 14, 2017, the Federal Communications Commission (FCC) abolished Obama's neutral network policy and re-empowered telecom operators with control over broadband Internet access. For the former network

neutral "three ban", Prohibit traffic control, prohibit payment priority manda-
tory broadband access behavior, the FCC said no longer supervise the future,
but requires businesses to strengthen self-regulation, such as the above-
mentioned behavior, only to the general consumer and regulators to increase
their own transparency can be.

The decree aroused the concern of all parties, the U.S. telecommunications industry applauded the neutrality repeal order, but the Internet industry
expressed dissatisfaction. In fact, the abolition of the network neutrality
policy is only a silhouette of the FCC to adjust the regulatory thinking of
the telecommunications industry, accompanied by changes in the direction
of a series of core policies, such as the classification of telecommunications
services, the FCC on its own power reform and many other issues. In fact,
since the presidential election in early 2017 and the Republican Party came
to power, the telecommunications industry has been making good progress.
This neutral network policy was repealed and it can also be regarded as the
official opening of the Light-touch Regulation era in the United States.

8.2.4 History of NN

The phrase of NN was first proposed in a law review article [7] by Tim WU in
2003, introduced the concepts of freedom, competition and innovation. NN
was only a concept suggesting that each network protocol layer should be
independent and perform the assigned duties at the original phase. With the
expansion of the commercial-scale internet, the focus of the competition has
been shifting from the connection and network layer to the application and
content layer which made the debate of NN switch from technological field to
commercial field. Because the term holds different meanings to people in different field, meanwhile NN has different debate focuses in different process
phases. Although NN is abolished, however, the debating will continue.

The 2015 FCC Open Internet Order raised a new round of debate.
Through the key event happened in U.S as follows:

- In December 2010, The FCC approved the Open Internet Order which
  was consisted of three items of NN regulations, and they are *"Trans-
  parency, No Blocking and No unreasonable discrimination"* [8].
- In January 2014, United States Court of Appeals for the District of
  Columbia Circuit (D.C. Circuit) overturned the Open Internet Order.
- In February 2015, The FCC issued Open Internet Rules and Order. And
  In June 2015, The Open Internet Rules and Order came into effect
  officially.

- In August 2015, The D.C Circuit announced the Open Internet Rules and Order will face an important federal-court test;
- On December 14, 2017, the Federal Communications Commission (FCC) abolished Obama's neutral network policy and re-empowered telecom operators with control over broadband Internet access.

8.2.5 The Analysis of Abolition of NN

On December 14, 2017, the FCC Committee voted 3:2 to repeal "Internet Open Regulations". The main contents include: First, abolish network neutrality policy, relax regulation, restore freedom of broadband access market. The FCC states that a network-neutral policy of "heavy regulation" of broadband Internet access services (BIAS) will increase the potential operating costs of the entire Internet ecosystem. Therefore, the FCC will regain the "light regulatory" approach 20 years ago to Stimulate growth and restore market liberalization and freedom. Followed by the business classification adjustment. The BIAS business is re-classified under Chapter "Information Services" under Chapter "Telecommunications Business" of the "Communications Act of 1934".

This adjustment will bring about four changes. First, the FCC no longer regulates the BIAS business, including both mobile and fixed broadband access. Secondly, the broadband access services launched by some Internet giants in the United States, such as the Google Fiber business conducted by Google, were previously included in the regulation of "telecom services" (and even triggered a "whether Google Inc. is an Internet company or a telecommunications company Enterprise" argument). After the current round of adjustments, the return of such businesses carried out by Internet companies falls into the category of information services "in Chapter 1 of the "Communications Act of 1934". Third, the power limit and part of the power out. The FCC banned state regulators from legislating for the network neutrality or issuing policies. In addition, the power of broadband consumer protection, Internet data security protection and other powers transferred to the Federal Trade Commission (FTC). Finally, ISPs are required to be self-disciplined, increase the transparency of information disclosure and require disclosure to consumers and government regulators of how they conduct paid-for-network access services.

According to the "Executive Decree on Rescheduling Internet-Related Policies at the End of November 2017" (WC Docket No. 17–108), the FCC has decided to abrogate the main reason for network neutrality. First of all, the basic attributes of the network access service are changed. FCC believes that the broadband access service provided by telecom operators should be a valuable commodity rather than a public product and should not be regarded as a public service such as water, electricity and gas, and should not be used for public service undertakings "Heavy regulation" approach. Therefore, in theory, telecom operators have the right to arrange their own business operations, the government should not be supervised. Second, the enthusiasm of telecom companies was hit and broadband investment declined. Since the introduction of the network-neutral policy by the FCC in February 2015, U.S. investment in broadband infrastructure has dropped for the first time since 2009, while the total investment in broadband investment in ISPs in mainland China dropped 3% and 2% year-on-year respectively in 2015 and 2016, while that of AT & T, Verizon, Comcast, represented by the top 8 US carriers in 2015 and 2016, the total investment in broadband fell 5.6%. For such results, the United States industry, economics and FCC believe that the network neutral policy is to combat the direct reason for the investment enthusiasm of telecom operators.

8.2.6 Current Discussion of NN

From the history of Network Neutrality, we can find that NN has been really controversial. Although US abolish NN, The essential debate will continue and the focus point will be discussed below:

8.2.6.1 Service innovation

The proponents of NN are mainly those enterprises that are related to Internet contents. Worriedly, they are stating that actions departing from NN principles could threaten the innovation of the internet content as, service providers may increase control on the content and applications over the internet. In order to encourage the innovative services with enhance quality of service especially from startups, the new EU net neutrality rules *"enable the provision of specialized or innovative services on condition that they do not harm the open internet access. These services use the internet protocol and the same access network but require a significant improvement in quality or the possibility to guarantee some technical requirements to their end-users that cannot be ensured in the best effort open internet* [9]. *These specialized or innovative services have to be optimized for specific content, applications or services, and the optimization must be objectively necessary to meet service requirements for specific levels of quality that are not assured by the internet access services*". The rules also urges these services cannot be a substitute to internet access service, can only be provided if there is sufficient network capacity and must not be to the detriment of the availability or general quality of internet access service for end-users.

FCC's rules refer above mentioned service as non-Broadband Internet Access Service [1] "*(non-BIAS) data services, which are not subject to the rules. According to the rules, non BIAS data services are not used to reach large parts of the Internet., not a generic platform—but rather a specific "application level" service, and use some form of network management to isolate the capacity used by these services from that used by broadband Internet access services.*"

8.2.6.2 The investment on the network infrastructure

The opponent of NN which is typically network operators argue that NN regulation will make it more difficult for Internet service providers (ISPs) and other network operators to recoup their investments in broadband networks, weaken the incentives to invest and upgrade the telecom infrastructures. Some network provider has argued that they will have no incentive to make large investments to develop advanced fibre-optic networks if they are prohibited from charging higher preferred access fees to companies that wish to take advantage of the expanded capabilities of such networks [10]. FCC reclassified the BIAS as telecommunication service in the Open internet Order [1], FCC believed that the reclassification will preserve investment incentives.

8.2.6.3 The management of internet traffic by internet service providers and what constitutes reasonable traffic management

Commonly, "traffic management is used to effectively protect the security and integrity of networks. It helps deal with temporary or exceptional congestion or to give effect to a legislative provision or court order. It is also essential for the certain time-sensitive service such as voice communications or video conferencing that may require prioritization of traffic for better quality. But there is a fragile balance between ensuring the openness of the Internet and the reasonable and responsible use of traffic management by ISPs [11]. The opponent of NN argued that NN may prove ineffective in such a dynamic framework nowadays, leading to welfare-loss caused by congestion problems, arguing in favor of the possibility of differentiation of data packets according to their quality sensitivity [1].

EU urged that all traffic be treated equally but allow the network providers to make reasonable traffic management in consideration of justified technical

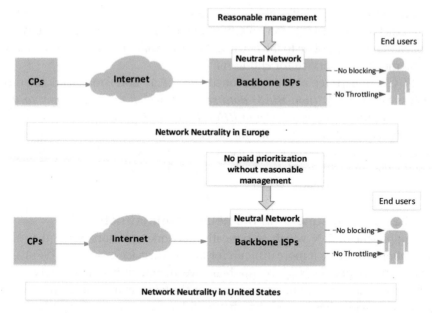

Figure 8.3 The comparison of NN between the US and the EU [8].

requirements, so as to preserve the security and integrity of the network or to minimize temporary or exceptional network congestion. According to FCC rules, the no-blocking rule, the no-throttling rule, and the no-unreasonable interference/disadvantage standard will be subject to reasonable network management for both fixed and mobile providers of broadband Internet access service. Figure 8.3 shows that EU comparison with the United States rules. We can see that regulators paid attention to the innovation when making their own NN policies. They leave room for the innovative services together with strict constraints.

8.3 CONASENSE [3]

In the past 10 years, wireless and Internet technology and system have created an explosive growth in personal and group applications which offered a wide range of data services. The limited data transmission capacity, as the bottleneck in the beginning, was broken. We can get much higher data rates over the same wireless channel and thus for support of many more demanding services. At the same time, we observe a rapidly increasing demand and innovative application areas for services related to positioning, tracking and navigation of some users/platforms. Similarly, an unprecedented development in sensing technology, sensors and sensor networks is being observed. A variety of sensors types are now available on the market in many domains. The integrated provision of these services will obviously raise people's living efficiency and Quality of Life (QoL). But traditional approaches may not be optimal, because of the allocation of different frequency bands, waveforms and hence different receiver platforms for these services.

Most promising CONASENSE services should be available in 5–20 years. The new CONASENSE services should reflect the trend toward an information society in which applications and services become equally important, bearing in mind that computing and communications should be integrated so as to save energy, software defined radio combined with cognitive radio concepts become increasingly important for new developments.

To achieve this goal, it is critical to identify the requirements for energy, terminal/platform and receiver/system design concerning diverse application areas including e-health, security/emergency services, traffic management and control, environmental monitoring and protection, and smart power grid. Minimization of energy consumption and energy harvesting deserve special attention in the novel CONASENSE architecture design mainly because of requirements for mobility, high data rate communications and signal processing and green communications. The novel CONASENSE architecture design will address problems and drawbacks of the existing infrastructures/architectures and be sufficiently flexible for future/potential developments. Consequently, the design of the CONASENSE architecture will be carried out so as not only to integrate existing and novel communications, navigation and sensing services but also to provide smooth transition between existing and new systems in hardware and software. NN impact on CONASENSE service innovation will be discussed in the next section.

8.4 Network Neutrality Impact on CONASENSE Service Innovation

From technology aspects, no discrimination requirement of Network Neutrality may have some negative impact on the new service innovation. The network nowadays is not neutral (Differentiated Network), it can supply quality of service (QoS) to different applications according to their characters and requirements. But according to the new NN rules, QoS measures will be taken as discrimination, and be banned in the pure neutral network, as shown in Figure 8.4. It is obvious to lower the network efficiency and will cause congestion easily.

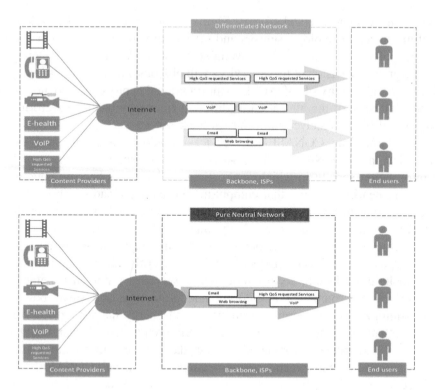

Figure 8.4   The difference between Differentiated Network and pure Neutral Network.

In the short term run, NN will certainly make Internet content companies have more innovation space, encourage more innovative applications and increase the efficiency of the society. But in the long run, NN will inevitably weaken the enthusiasm of the investment on the network construction, lower the network quality level gradually. This in turn will influence Internet companies in the end. The reasons are as follows:

(1) From the technical aspect, NN will impact the quality of the Internet service because NN limits the ability to guarantee the network QoS. As of now, there are mainly two types of model of QoS [12].

- Integrated Services (IntServ) uses RSVP (Resource Reservation Protocol) for signaling to invoke a pre-reservation network resource and traffic handling. IntServ can provide end to end guarantee for services and applications. But because Intserv is expensive and time-consuming, it has not been widely used in the Internet.

- Differentiated service (Diffserv) is a mechanism that identifies and classifies traffic in order to determine the appropriate traffic handling mechanism. It can integrate the same type of services and manage them together. It is now used widely.

To exemplify the models of QoS, we can imagine an accident site, there are several wounded persons: some of them got severely injured and need to go to hospital immediately, while others are not so urgent. When the ambulances come, the doctors do not do any diagnosis and ask the severely wounded to get on the ambulance which makes some non-significantly wounded cannot get on the ambulance in time. This case is Best-Effort Model (no QoS). The IntServ model is that the wounded whoever is serious or not need to reserve the ambulance, if the severely wounded didn't make a reservation, he will not be sent to hospital. In the Diffserv model when the ambulances come, the doctors will distinguish the level of wounded and ask serious wounded persons with similar situation to get on and be treated. So we can find Diffserv is an effective method for application and service transmission on Internet. But with NN, differentiated service is banned in public network.

(2) Different Internet services and applications have different requests. QoS uses four parameters to judge the services request including bandwidth, delay, delay variation and packet loss rate [13]. Table 8.1 presents the definition and the impact of the four parameters of QoS.

Table 8.2 shows that some services are sensitive to delay but not to packet loss, and some services are sensitive to packet loss but not to delay. It is necessary for a network to treat packets belonging to different applications

Table 8.1 Definition and Impact of the four QoS parameters [13]

Parameter	Definition	Impact on QoS
Bandwidth	The maximum data that can be transmitted per second in the network.	It is to measure the transmission capacity of the network
Delay	The transmission time a service packet takes from one nodes to the other one	If the delay is too long, it will lower the QoS
Delay variation	It means the variation of the packets delays in the same flows	It is a key factor impacting QoS
Packet loss rate	It means the rate of the data packets that lost in transmission	Low packet loss rate will not impact the QoS

Table 8.2   QoS Parameters for Some Applications [14]

Type	Bandwidth	Delay	Delay Variation	Packet Loss Rate
Email	Low	Not sensitive	Not sensitive	Not sensitive
WhatsAPP	Medium	Medium sensitive	Medium sensitive	Not sensitive
Video application	High	Sensitive	Sensitive	Sensitive
E-commerce	Medium	Sensitive	Sensitive	Sensitive
IoT, Industry 4.0	High	Sensitive	Sensitive	Sensitive
CONASENSE services	Large	Sensitive	Sensitive	May not very sensitive

differently in terms of their different requests. For instance, a network should give low-delay service to VoIP packets, but best-efforts service to e-mail packets [14]. Different services also need different QoS provisioning.

8.4.1 From Social and Economic Aspect

NN may reduce the incentive of investment on the network construction. As we know, the funds usually flow to the market with big profits. If there is not enough profit from telecom network, investment on the network will certainly be reduced. Internet companies may also lose their interest to improve the efficiency of transmission, because they do not need to pay for the overuse of bandwidth. Insufficient investment on network with the overuse of bandwidth will certainly result in network congestion and inefficiency.

8.4.2 The Influence of NN Abolition

The abolition of neutrality of this network will have an impact on the telecom and Internet industries respectively. Firstly, the abolition benefits the telecom industry. The telecom operators (broadband service providers) will see significant positive changes. The enthusiasm for broadband infrastructure investment is expected to be significantly improved, and the "digital divide" in remote areas is expected to be resolved again. Specific to Comcast, Verizon and AT&T and other traditional telecom giants' rights and interests, on the one hand, the network can be opened up a separate "fast lane", according to the quality of transmission services to the Internet companies charge varying levels of transmission costs. On the other hand, the priority delivery of its own content business, such as Comcast For its own large media group NBC

Universal, Comcast can give priority to the delivery of NBC's business content. For specific content, specific software and services, telecom operators have the right to limit.

Followed by the Internet industry will be polarized. After the abolition of the neutral network policy, the U.S. Internet industry expressed its protest to the FCC in its view that abolishing network neutrality is equivalent to giving telecom operators the power to manipulate Internet traffic and discriminating against Internet businesses, which will eventually aggravate the polarization of the Internet industry. First of all, for Internet giants such as Google Facebook and Microsoft, they will be forced to pay the telecom companies what they call "fast track" fees and buy high-quality services. However, due to the huge financial resources of the Internet giant, it is expected that there will not be too many negative effects. Second, for the Internet small businesses, the network transmission quality of startups may decline due to the lack of protection of network-neutral policies, transmission costs may increase, and the long run will affect network innovation and business growth. Precisely because of this, the US neutral network supporters that: "FCC abolition of network neutrality policy, not to destroy Google, but the next Google".

The neutrality of FCC legislation in the United States has lasted for more than 20 years. The difficulties and hiccups among them can be imagined. The adjustment of this network neutral policy is, in essence, the rebalancing of the future development space between the telecommunications and internet industries by the U.S. telecommunications regulatory agencies Caused. During the Obama administration, telecom companies invested a great deal in advancing the network investment and Internet companies enjoyed too much dividends. After taking office, Trump passed a neutral policy adjustment and handed over a part of bonuses to telecommunications companies. The FCC's stance reversal is compensatory from a deeper point of view.

In the United States, network neutrality will still be the focus of controversy. Comcast, the United States, Verizon and Google, YouTube fierce competition, but on the other hand, the two sides can take a cooperative approach, the Internet companies own innovation and development is the need for telecom operators to support In the future, the cooperation between the two industries is still greater than the competition. In the long run, the surge in broadband network traffic, the consumption of operators such as video and other services, the competition between basic operators and Internet companies, the upgrading of networks and the sustainable development of the market also exist. In addition, taking into account the further integration of telecommunications and the Internet, although there will be new opportunities for development in this process, conflicts and conflicts are unavoidable in the process of concrete integration.

8.4.3 Impact on CONASENSE Services

The large amount of CONASENSE services, to some extent, can be taken as environment-sensitive services. And for people on air, on vessel, or on car, on road, their environmental situation including location, speed, temperature, health condition etc. are constantly changing. All the relevant information should be received correctly and timely for a CONASENSE services will probably use them to make a correct and in-time decision, to help improve people's Quality of Life (QoL). NN rules put relatively strict criteria of specialized services or no-BIAS services, sometimes on a case by case base. That may slow down the CONASENSE services experimenting, developing, commercializing process.

High data transmission capacity is the base for the CONASENSE services. Most of the CONASENSE services focus on the future, on the assumption of WISDOM or 6G network deployment and extensive use of sensors. The value of CONASENSE services will not be realized on a congestive and no guarantee networks.

8.5 Conclusions

NN rules, its relevant debates, and, its impact on ISPs and ICPs are discussed in the Section 8.2 of this chapter as a background study. In this discussion, NN rules such as NB, NT, and NPP were covered including their exemplified definitions and impacts on present communication paradigm. It is discussed that these rules are user centric and may hamper the financial benefit of ISPs. The absence of any control in the flow of information through a communication network may refrain ISPs from earning quality based revenues. In such a case, ISPs may not be willing to enhance or upgrade the network to improve the QoS of the network. Therefore, although NN rules are beneficial for innovations, as campaigned by NN proponents, provided by NN rules, the end user may not "really" enjoy the QoS.

With the convergence of Internet and telecommunication network, basic telecommunication services have been moved to the public Internet network. Many of them has quite strict QoS requirement so as to guarantee the relevant service responsibilities, such as emergency call, the basic service quality level. In this process, the policy maker and regulator should be careful to handle between service innovation, lawful customer right and social responsibility. QoS models have enough reasons to remain as legal network functions. And currently, US has realized the problem, and decide to abolish NN.

In the Section 8.4, it is discussed how NN rules, in their present form, will impact CONASENSE service in the future. Through some assumed futuristic scenarios it is discussed that NN rules may not be favorable for innovation service. Banning Diffserv may put innovation service in the category of a usual communication system and ignore the very high data demand of the CONASENSE service. As customers may not welcome the innovative but poorly served new technology, the aspirant companies will struggle capturing market for this innovative technology. Further, NPP may not allow users to choose better services among the choices offered by ISPs.

NN policy may stimulate the development of service innovation in short term but may be not good for the network base in long term, which may in turn hinder the service innovation, such as CONASENSE services. The relatively strict criteria for specialized services or no-BIAS services may slow down the CONASENSE services experimenting, developing, commercializing process.

Regulators are suggested to make careful decisions on NN policy to guarantee the long term benefit according to their own situation. All the stakeholders should be encouraged to find a way to ensure the prosperous development of the industry in the market on their own, especially in WISDOM era [15].

References

[1] FCC, "The Open Internet Rules and Order" FCC 15–24, March, 2015.
[2] FCC, Category "For Consumers " "Open Internet", 2015.
[3] Cory Janssen, "Over the Top Application (OTT)," Techopedia http://www.techopedia.com/definition/29145/over-the-top-application-ott, 2015
[4] Leo Ligthart, Ramjee Prasad, "Communication, Navigation, Sensing, and Services (CONASENSE)" River Publisher, 2014.
[5] Winston Maxwell, Mark Parsons, Michele Farquhar, Net Neutrality – A Global Debate, Hogan Lovells Global Media and Communications Quarterly 2015, P15–17.
[6] European Paralimentary "Our commitment to Net Neutrality", EU Actions, October, 2015.

[7] Tim Wu, Network Neutrality, Broadband Discrimination, Journal of Telecommunications and High Technology Law, Vol. 2, p. 141, 2003.

[8] FCC, "The Open Internet Order" FCC 10–201, December 21, 2010.

[9] European Commission,, "Net Neutrality challenges", October, 27, 2015.

[10] Dong-Hee Shin & Tae-Yang Kim, "A Web of Stakeholders and Debates in the Network Neutrality Policy: A Case Study of Network Neutrality in Korea".

[11] European Commission, "Roaming charges and open Internet: questions and answers" 27 October 2015.

[12] El-Bahlul Fgee, Jason D. Kenney, William J. Phillips, William Robertson and S. Sivakumar, "Comparison of QoS performance between IPv6 QoS management model and IntServ and DiffServ QoS models" the 3rd Annual Communication Networks and Services Research Conference, 0-7695-2333-1/05, 2005, IEEE.

[13] Fabricio Carvalho de Gouveia and Thomas Magedanz "Quality of Service in Telecommunication networks", Telecommunication systems and technologies, Vol, II.

[14] Hua wei technologies co(2013), ltd, "QoS Technology White Chapter", http://e.huawei.com/us/marketing-material/onLineView?MaterialID=%7B3623FE01-3572-4413-A71B-EBEBE9F2E141%7D

[15] Ramjee Prasad, 5G Revolution Through WISDOM, Springer Science+Business Media New York 2015, Wireless Pers Commun(2015)81: 1351–1357 2015 March.

Biography

Yapeng Wang obtained Master Degree from Beijing University of Post and Telecommunication in 2008. Currently she is a guest researcher in CTIF in the field of Network Neutrality. Before 2008, she worked in Teleinfor Institute of CATR as a researcher and the field is telecommunication regulation, policy and market. Till now, worked in International Cooperation Department of CATR, and responsible for EU projects and ITU-D issues in China. The projects include:

- 2011–2012 Promotion of Green Economic Growth by Broadband Network
- 2012–2013 Open China ICT Project–Observation of the Chinese telecom' development
- 2012–2013 Implementing and planning outline of 'Smart Qianhai's policy (Qianhai is a region of Shenzhen)
- 2012–2014 International standard assessment of China Unicom from 2010 to 2010
- 2014 WTDC: China Reception, Editorial Committee, Election for Mr. ZHAO Houlin
- 2014 PP: China Reception, Editorial Committee, Election for Mr. ZHAO Houlin
- 2014 Application for ITU Centers of Excellence (CoE) in Conformance and Interoperability (C&I)
- 2014 Application for Conformance and Interoperability Test lab
- 2015

Publication

- The Network Neutrality impact on the OTT Service, Yapeng WANG, Ramjee Prasad, ITU-D Study Group 1
- The Network Neutrality impact on the OTT Service of 5G WISDOM, Yapeng WANG, Ramjee Prasad, International Symposium on Wireless Personal Multimedia Communications
- The impact of Network Neutrality on HBC, Wireless Personal Communication Journal, Yapeng WANG, Ramjee Prasad, (88(1), 97–109, doi: 10.1007/s11277–016–3245–5)
- Network Neutrality for CONASENSE Innovation Era, Yapeng WANG, Ramjee Prasad, CONASENSE Book (accepted)

9

"The Vitual Business Model" – Towards a Virtualized Communications World – Challenges and Opportunities

Peter Lindgren

CGC, Business Development and Technology, Aarhus University,
Aarhus, Denmark

Abstract

A variety of challenges and opportunities are brought to our world and its businesses these days, especially when business and business communications turns virtual and thereby business models also are transformed to or created as virtual business models (VBM). But

What does it actually and really mean that a Business Model is or becomes virtual and communicates virtually?

And

What does it actually and really mean that a Virtual Business Model communicate and converge with the physical and digital BM Ecosystem?

The ability to develop and understand Virtual Business Models has become a very interesting topic. The cornerstone of future business and competitiveness might lay in these different types of virtual Business Model's.

However firstly we have to understand better the term "business models" and then Virtual Business Models.

To achieve this the chapter takes its point of entry in an earlier defined multi business model approach developed by the CGC/Mbit research group and combine this with different definitions of virtual and virtuality. The chapter use this as "a stepping stone" combined with several case examples of Virtual Business Models.

Towards Future Technologies for Business Ecosystem Innovation, 137–156.

The objective of this chapter is to report on studies carried out on Virtual Business Model's. Different Virtual Business Model cases are presented. The aim of this work is threefolds.

1. Firstly to support the development and journey of defining a conceptual model of Virtual Business Models initiated and to some extend presented at the round table speech at the Global Wireless Summit 2017 IEEE conference in Cape Town, South Africa.
2. Secondly to show how the VBM framework model and concept can be adapted by businesses aiming at or already using Virtual Business Model and Virtualizations techniques.
3. Finally the chapters aims at proposing a generic language for VBM to make it easier for businesses, academians and others to talk, understand and innovate VBM.

Keywords: Virtual Business models, Virtual Communication, Business Models, Virtual Business Model innovation, Virtual Business Model Ecosystems, Sensors, Wireless sensors.

9.1 Introduction

Not many have defined what a virtual business model (VBM) is. To understand the term "Virtual Business Model" the chapter take its point of entry in the multi business model approach and business model (BM) terminology developed by the CGC/Mbit research group [4–6] and then use this as "a stepping stone" to try to explain how a VBM could be defined. Some of the basic BM terminologies used in the chapter are referred to in our previous chapter – the business model cube [6] forming a generic language for how to define a business model.

The author took the freedom to change the terms – company, enterprise, firm and organization – in registered literature and academic work on VBM and use the generic term business [6] as the umbrella term for these above mentioned terms. When ever this was done it was marked in *italics* as *business* in the original authors text. Hereby it was easier to define and characterize what the authors on VBM really were talking about and addressing – and where their contribution could be placed in the levels on business model innovation as presented in Figure 9.1.

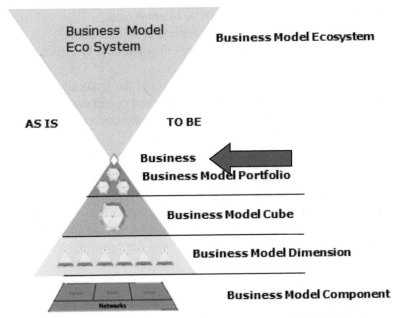

Figure 9.1 Business Model Innovation Levels – a conceptual model for business model innovation layers adapted from Lindgren and Horn Rasmussen 2013 [6].

9.1.1 Virtual Related to the Levels of the Business Model and Business Model Innovation

A literature study shows that not many have tried to define a VBM. Bryder, Malmborg-Hager and Søderlind [7] tried to do this by defining a Virtual business model (VBM) as

> "a way to organize an innovative startup *business* and facilitates increased flexibility in the use of both financial and human resources".

She further claims that a VBM can "promote development of new ideas and inventions". She however also point to the fact that "a VBM should not be confused with Virtual business".

In the virtual *business*, the utilization of the financial resources can be optimized with value adding, cost-effective production and change or new business model as a result. This business model – a VBM – is defined using several criteria;

- the *"business"* has a limited number of employees;
- the management has competence for business innovation;
- the *"business"* has financial resources to perform or has the ambition to find such financial resources;
- the *business* has a defined plan for the use of the financial resources;
- the majority of the *value chain functions* and operations are carried out and performed by *networks* (called External Resources Provider) outside the Virtual *business*;
- the ownership of the created value (e.g. technical results, patents) developed by the *networks* belongs to the virtual *business*" [7, 8].

After the foundation of the virtual *business* and the development of the business using external resource providers, the *business* can continue to use the virtual *business* format for continued business model innovation (BMI) or after some time the *business* can transform the business to a traditional integrated *business* [7, 8].

9.1.2 Virtual Related to the Levels of the Business Model and Business Model Innovation

The work of Bryder, Malmborg-Hager and Søderlind [8] leaves us however with some unanswered questions and some considerations related to the Figure 9.1 on the Business Model Innovation levels. The authors seems actually referring to primarily "the *business* layer" as we see it. To make it a little more clear on where do the authors talk about the virtual BM topic, we indicate in Figure 9.1 with the blue arrow where we see Bryder et all focus their VBM definitions and major part of their work.

Firstly it can be seen from the text that the authors [] defines the VBM as on the Business level in the Figure 9.1 and as a business that is newly established – a startup business – very close to "To Be" Business level – as also indicated with the blue arrow. This means that the term the authors use in their definition of virtual is related to the overall business BM level – and fragmentally a startup business. Needless to say but we think this is a little to narrow definition and understanding of a VBM.

This definition and understanding can however also be rather radically as it could mean that all 7 dimensions [6] of the overall business BM's are understood as virtual or it could opposite mean that it is meant incremental related to virtuallity as just some of the overall business BM's dimension could be or are actually classified as virtual. It seems however from the authors side [7] that their definitions just address some of the BM dimensions as virtual – e.g. the network BM dimension. This - we believe and show in the paragraph beneath – is again a rather narrow understanding and classification of a VBM.

What is meant by that is that the overall business BM could potentially be classified as virtual but the other BMI layers could actually operate more or less virtual. The opposite could however also be the case – and any mixture of this could be the case.

This is however not explained very clearly in the text [7]. However it seems as if the examples they use indicate that only some BM dimensions of the overall business model are classified as virtual – and the rest are left out. This makes it very difficult to use the authors terminology as a framework for measuring VBM's – and particularly the degree of VBM – How virtual is the BM.

We propose therefore on behalf of our findings that we will express in more details later in the chapter, that businesses are seldom pure virtual on the overall Business level as seen in Figure 9.1 – but are more virtual on the BM, BM portfolio, BM Dimension or BM Component level. However the business could also aim at another BMI level the Business Model Ecosystem Level [2]. – focusing on creating, capturing and delivering a Business Model Ecosystem that are Virtual – but this does not mean that the Business needs and are forced to be virtual to fulfill this strategy.

The last part leaves us with a minor note and challenge related to the term virtual – that the business overall BM maybe is not strategically aligned on its operation and/or intended operational on virtuality. In other words – it might not stick to its core overall business model – and thereby may not stick to the business core business e.g. core business competences [9]. This could lead to weaker business results according to Mckinsey survey [10] and Prahalad and Hammels [9] earlier observations.

"Sticking to the core business" means – idealized – that the business overall business model (Figure 9.1) is constructed and intended to express that the business "main" or "core" value propositions, user and customers, value chain functions, competences, network, value formula and relations [6] are aligned and any in the business overall BM are chosen or rejected according to this "ambrella" of the overall BM of the Business.

However, as we can see from the authors example [7] this is not the case in the authors definition of a VBM as they address the virtual to specific parts of the BM – the network. This seems to be restricting the real opportunity of a VBM, as a VBM strength lays in its ability to link more dimensions of the BM e.g. technology, human resources, organizational systems and

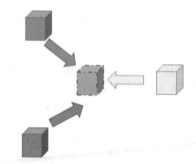

Figure 9.2   Creating a Virtual Business Model.

culture virtually in the aim of pulling opportunities to the businesses and its BM's involved. This view and approach is highly proposed and inspired by Goldmann, Nagel and Preiss [12] as they argue that

> "A virtual *business* is an Business opportunity pulled and Business opportunity-defined integration of core competencies distributed among a number of real *businesses*" [12].

The core competencies we will comment on later, but according to this definition the virtual business is equal to an integration of different real (physical) businesses (the business level Figure 9.1.) gather around an defined opportunity – as we define it a "TO BE BM" – where they aim at creating, capturing, delivering a business model on behalf of more businesses BM's.

More precise the businesses are gathered in the aim of integrating some dimensions of one or more of their "AS IS" BM's and/or create some new BM Dimensions together in the aim of creating a "TO BE" VBM together as shown in Figure 9.2.

9.1.3 Virtual BM's Related to Competence, Capability and Core Competence

Again as we stressed above the change BM and/or "TO BE" BM can be more or less virtual on the seven dimensions of the BM. This means that we are able to measure the virtuality related to the BM dimensions related to the number of BM dimensions change to become Virtual. This we propose called the *complexity* dimension related to how virtual are the BM's dimensions.

Business are claimed to have a limited number of employees which makes it optimal these days to create and look for VBM providing access to qualified competences. These businesses BM's must have competence for creating, capturing and delivering value propositions (product, service and process) virtually and thereby doing virtual business model innovation often within high speed – the time perspective of a VBM. As we found in many of our cases several VBM's do not exist or are not present for a very long time. Hereafter the businesses behind the VBM are often vanished and very hard often impossible to track.

Many authors claim that "VBM *business*" must have or have access to financial resources to create, capture and deliver these VBM. They must also have ambition to find financial resources if the VBM is in lack of financial resources. The business involved must also be able to fast define a plan for the use of the financial resources; and the majority of the operations of VBM requires Businesses that have the capability to execute the VBM [7].

In such VBM case the shared competences and the ability to share and "pool" competences fast in a VBM can turn up to be a core competence [9] – maybe the core competence of a VBM. As this can fulfill the criteria to be a core competence – 1. be unique and 2. differentiate itself to other BM proposed in the BMES and 3. it can be used to give competitive advantage to the business involved. Herein lays maybe one of the characteristics that differentiate a VBM to other BM's.

However opposite to previous definitions on core competences the VBM is not built on and around a single Business BM – but instead is "a merge" of different businesses BM's. This means that the VBM if it fulfills the criteria of a core competence [9] is based on a shared core competence. This makes it very difficult to measure, very difficult to track and compete.

9.1.4 Virtual BM's Related to Network and Network Based BM's

In our investigation it was noticed that many authors [7, 8, 12] claim that VBM's are built on networks – in other word they believe that VBM are special as they are network based business model. Our study shows very clearly that all BM's are network based – the network BM dimension – and VBM can not be claimed to be special on the network dimensions – as it would never function as all other BM's without involvement of network partners. In Figure 9.2 we made a sketch of a the VBM and we propose that a VBM as all other BM's is defined as a network based business model (NBBM) [15], which in our terminology all business model actually are [6]. There seems on this point therefore to be no difference between a "normal" BM and a VBM. Therefore we propose to also look for some other characteristics to distinguish a VBM to any other BM's In this case we propose to look deeper into the term virtual and what virtual actual means.

9.1.5 What Does Virtual Actual Mean – and How Has it Previously Been Defined?

The word Virtus – [femininum] – according to the Latin dictionaries [13, 14] means

1. Strength and Power
2. Courage and bravery
3. Worth – manliness – virtue – character – excellence
4. Army
5. Host
6. Mighty works
7. "A class of angels"

> "The adjective Virtual was describe in Late Middle English (also in the sense 'possessing certain virtues'): from medieval Latin virtualis, from Latin virtus 'virtue', suggested by late Latin virtuosus" [14].

The adjective virtual was used to describe

> "something that exists in essence but not in actuality" [13].

In other words something – in this case a BM – that "exists in essence but not in actuality". How can this be? and How can this happen?

To come a little closer to this many have been inspired by and relate VBM to the computer and Information, Communication Technologies (ICT) – having said that virtual is related to these or – as some claim can only be realized by computers and ICT.

Although virtual can be used to describe anything that exists in effect, but not in fact, it is often used to relate to and describe things created "in a computer", "online world" or "virtual worlds" like Second Life, World of Warcraft e.g.

> "She enjoyed playing the virtual role-playing game with her online friends".

An "item" provided by these computers and their embedded software may sometimes be described as being a virtual item, when it is a representation or non-tangible abstraction of the physical object, or it is a functional emulation or simulation of it. Hereby the field Artificial reality and Virtual Reality

Figure 9.3 Business Modellist playing with a Virtual Business Model with Virtual Reality Technology.

together with AI and Virtual Reality Technology as seen comes into play (Figure 9.3).

Virtual reality (virtuality), computer programs with an interface that gives the user the impression that they are physically inside a simulated space – "a place" where they can "play" or innovate with e.g. BM's takes BM's and BMI to a new dimension. It makes it possible to see the real world in a different perspective but also to be independent of the physical worlds constrains.

"Virtual memory, a memory management technique that abstracts the memory address space in a computer, allowing each process to have a dedicated address space" have in this case been a very important tool in the ICT BMES to increase the amount of storage of data.

The term virtual has been as can be seen in Bryders [7] definitions very much been related to computers and ICT as the basis for VBM. "Virtual World, a computer-based simulated environment populated by many users who can create a personal avatar, and simultaneously and independently explore the virtual world, participate in its activities and communicate with others have also been an important initiative to create VBM – but it has still some way to go, when fully accepted.

9.1.6 Virtual Businesses

"*Businesses* that conduct their business activities solely online" [16, 17] has as can be seen above often been classified as a VBM. However, we find in literature [12] and in many of our business cases that VBM can also be created as a physical VBM or as a combination of Digital and Physical BM's.

Summing on the above-mentioned discussion we propose therefore that

> "a VBM can be measured as both a physical, digital or a combination of a physical and digital BM"

that

> a VBM can have different degrees of virtuallity, so that "a VBM can have one or more over all BM dimensions that are virtual".

Further we propose that a Virtual Business can have

> "BM's that are virtual – and/or even BM's that have one or more BM dimensions that are virtual and even again some BM components in a specific BM dimension that can be virtual".

In other words we propose that the degree of virtuality of a VBM can be measured on different Business Model Innovation (BMI)'s layers as indicated in Figure 9.1.

Hereby we open up to a much broader, flexible and even some would say more agile definition of VBM and the VBM approach and terminology.

As we find it very difficult to find many or even one Business at the overall Business Model level (Figure 9.1) that is completely virtual as defined above – the model above seems to be a way to measure virtuallity of a BM related to the BM, BMES it is addressing and the BM process it is carrying out. Further as on overall BM portfolio, BM's, BM dimensions, BM Dimension components the model could also be increased and further be useful to give the business and somebody from outside that want to measure the business on virtuality an overview of what parts of the business innovation levels are virtual (Figure 9.1).

Having said this, we could imagine in this new terminology that a BM could theoretically and in the future be pure virtual on all BMI levels as seen in Figure 9.2. However we did not in our study find a pure VBM.

9.1.7 Virtual BM Related to How New the Business Model Is

The way that the previous mentioned authors define VBM as if VBM is related to how new the Business is or to a new business – we could not verify on behalf of our investigations and findings. Our findings show that

From Physical to digital to virtual Business Model, BMES and Process!

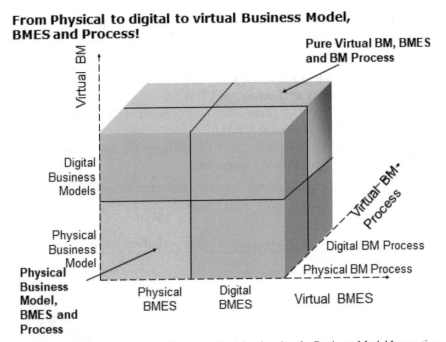

Figure 9.4 Degree of virtuality of Business Models related to the Business Model Innovation Layers – a conceptual model Lindgren 2018.

Virtuallity of a BM can be both with new as well as existing BM's. Therefore we propose that VBM has nothing to do with how new a Business or a BM is – as an old VBM can equally turn up as VBM at different times and vanish again after it has fulfilled its task.

In other words on behalf of our findings – as we will comment on later – we found that a VBM are related to both the **time** it "live" and **the task** it carries out or has to carry out. A VBM is in our option – time dependent – and will not be present at a certain time if there is no task for it.

The above mentioned manifold of definitions of VBM leaves us however still with one challenge as:

How do we really define virtual related to a business model?

9.1.8 How Do We Define Virtual Related to Business Model?

In our Multi Business Model approach and terminology [6] we have earlier distinguish between different overall Business Models. Using this terminology and transferring it to the area of VBM we generally, identified three types of VBM's which do not necessarily conflict but rather show the different characteristics a VBM can have and take. We propose on behalf of our research findings that the overall VBM types included:

1. The Physical VBM's – where most BM dimensions are Physical – and the Physical VBM are leading the VBM
2. The Digital VBM's – where most BM dimensions are Digital and the Digital VBM is leading the VBM
3. The Mix VBM – where the BM Dimensions are based on a mix of Physical and Digital VBM Dimension

In the case research, we identified 3 overall characteristics of VBM's as seen in the Table 9.1.

Table 9.1 Characteristics of physical, digital and virtual BM's on overall Business Level

Overall Virtual Business Model Types	Context Characteristics of VBM's	Example	Case Examples
Physical VBM's	VBM mainly based on physical value propositions, users- and customers, value chain functions, competences, network, value formula, relations. VBM turns up when there is a demand or task for it	Café on demand, Butcher on demand, Hairdresser on demand Phycologist on demand that are present, whenever it is necessary or there is a task for their BM. Coffee Van Ice Cream Van Food delivery service On demand Bicycle repair business	Waiter.com Justeat.com Carclass.com Easybike.dk

Table 9.1 Continued

Digital VBM's	Security Software and Update of software that	Rent and use of car, hotelroom, video/films and games serviced via a digital VBM	

UBER.com

AIRBNB.com

NETFLIX.com

World of Warcraft – Blizzard.com

The MIX VBM	VBM's created, captured, delivered, received and consumed by a mix of BM's with high degree of dynamic, flexibility, agility, where network partners BM's constantly comes in and goes out of the VBM's according to the task and demand of the BM.	An "AMBIA" – a business with many workers but with no employees	

Redcross.com

www.qq.com

SecondLife.com

Bitcoin

As can be seen the VBM can involved different combination and mixtures of the above mentioned VBM's types.

Table 9.2   Characteristics and business cases of different VBM in physical, digital and virtual BMES's

Type of Virtual Business Models	Physical BMES	Digital VBMES	Virtual BMES
A physical VBM	An Ice Cream shop seasonal or occasional opened A Food delivery Wagon A bike repair business that comes on demand (Easybike.dk) A bus business (Midttrafik) use physical busses in combination with electronic travel tickets (Rejsekortet) Midttrafik.dk	Containers with safety power system delivered with secure power machine delivered to run Digital VBM delivered for Japan at the tsunami/atomic accident (Deif.com)	E-sport event where participants turn up in a container delivered by a VBM at and esport event and plays e-sport in the tailor made container
A digital VBM	A supermarket uses an ecommerce site to have physical customers order goods – (Bilka.dk) Mobile payment for buying goods (Mobilepay.dk) Digital Wearable and security delivered to a Musik Concert – Monica EU project – (www.monica-project.eu)	A online e-shop opened for one event/conference Security software update on computers in a business delivered on demand or when update (Semantic.com)	Digital Virtual Business selling virtual clothing in second life
A Pure Virtual Business Model	Virtual Money payed in a Physical BMES for e.g. a Taxi Drive (Bitcoin)	Digital money payed in a digital BMES (Bitcoin)	Virtual clothing produced and delivered in a virtual world by virtual Business. Payment with virtual money (Linden Dollars) in a Virtual World (SecondLife.com)

It should be noted that it was very difficult for us to find a "pure" physical, digital or virtual VBM. Most VBM's are a mix of these types.

This leaves us finally with some context consideration related to the characteristics and definition of VBM's and related to the BMI layers in which we are discussing VBM and VBM innovation.

As can be seen in the table beneath we propose some basic characteristics to classify a VBM.

Table 9.3   VBM Characteristics and context

VBM Context and Characteristics	Questions to be Ask Related to VBM Context and Characteristics
Time	When is the VBM Virtual?
Task	What is the VBM proposing or going to propose?
BMES	Where are the VBM proposed and going to be propose?
Process/lifetime	Where in the BMI process (creation, capturing, delivering, receiving, consuming) are the VBM operating and going to be operating?
Sharing of Competences in VBM	How in the BMI process are the VBM competences shared?

9.1.9 Challenges and Potential of Virtual Related to Business Model?

The interest of Virtual Business Models are growing, because it is value increasing, cost saving and smarter for business, users, customers, network partners, technology and even employees involved compare to other BM's. Lots of gains and release of pains can be achieved.

However we verified in the chapter that there is a great fuzziness around the terminology and definitions of VBM. Several relate VBM to fragmented part of the BM's dimension – e.g. network. Further some related it to start up businesses and new BM's and network based BM's.

In the chapter we discussed why these definitions seems not to be enough to explain the difference between a "normal" BM and a VBM. The chapter proposed a different approach to VBM definition – as related to the Business Model Innovation levels and characteristics of the VMB and the context of where the VBM are operating or are intended to operate of the VBM.

Virtualization of Business Models is definitely an enhancement associated with the application of the integration of the Physical, digital and virtual Business Model.

New and smarter technology in all aspects enables increasingly BMI, BMI Process and Business Model's (BM) to become more virtual in all part of their lifetime. However as we argue VBM can be pure physical, pure digital and a combination of both.

This together with smart analytics [1] affect all areas of VBM's life and lead to new ways of working, communicating, operating and cooperating with BM's, in BMI processes and Business Model Ecosystem (BMES) [2].

For the business, it is about access to potential BMES (Physical, digital or virtual) and having an advantage by virtualizing their BM's, BMI and BM's operations. Businesses gains hereby greater efficiency, intelligence, value and agility. Businesses have always adapted with the changing times, but the influx of technologies that enables them to make a larger part of their Business Modesl virtual has accelerated the pace at which today's businesses need to evolve VBM's and the degree to which they transform the way they operate, innovate, and strategically manage VBM's. So each business can build its co-operational, and financial arrangements, manage its VBM's to achieve its strategic visions, missions, goals and strategic plans.

9.2 Conclusion

The Multi BM approach was used through our research on VBM [3]. The research addresses the gap in research on VBM and the strong demand to find a generic definition and language of VBM's. The significance and importance of this work is related to the huge unexplored possibilities that VBMI offers today, when we fully understand the BM levels, BM dimensions and BM dimension components of the VBMs thoroughly and are able to communicate, work and innovate with VBMs at all these levels.

Meanwhile, the ability to develop innovative VBMs fast, efficient and effective has become a growing cornerstone for increasing competitiveness and survival of many businesses. VBMI appears even more important as digitalization and virtualization of BM's are placing substantial stress on businesses. As a consequence, VBMI and the competence of VBMI have become more important for many business and its network – even to get access to BMES. Accordingly VBM and the process of VBMI can be described as a phenomena. Building on this, the VBM and VBMI as a conceptually distinct construct may provide BM theories with new explanatory power and reach to both the technology and Business Model Community. Following this trend, experts can be given tools to manage better their VBM's, VBMI and VBMI Processes.

One such tool can be the Virtual Business Model approach presented in this chapter but also professionalizing VBMI Leadership and management. It further opens the potential for new concepts and deeper understanding of VBM. The different kinds of networks apparently allow for various possibilities. The businesses can achieve several synergy effects by combining such different kinds of VBM's not least in relation to high speed Virtual Business Model Innovation (VBMI).

References

[1] Claudia Loebbeckea, Arnold Picotb. Reflections on societal and business model transformation arising from digitization and big data analytics: A research agenda, The Journal of Strategic Information System, Vol. 24, Issue 3, September 2015, pp. 149–157.

[2] Lindgren P, (2017). The Business Model Ecosystem Journal of Multi Business Model Innovation and Technology, Vol. 4(2), 61–110. doi: 10.13052/jmbmit2245-456X.421 River Publishers.

[3] Teece, D.J. (2010), "Business models, business strategy and innovation", Long Range Planning, Vol. 43 Nos. 2–3, pp. 172–94.

[4] Lindgren, Peter, Multi Business Model Innovations Towards 2050 and Beyond, Wireless World in 2050 and Beyond: A Window into the Future!. ed. / Ramjee Prasad; Sudhir Dixit. Springer, 2016. pp. 149–160 (Springer Series in Wireless Technology).

[5] Lindgren P., and Ole Ramusssen (2012). Business Model Innovation Leadership Journal of Multi Business Model Innovation and Technology, 53–69. River Publishers.

[6] Lindgren, P. and Horn Rasmussen, O. (2013). The Business Model Cube. Journal of Multi Business Innovation and Technology, 1(3), pp. 135–182.

[7] Bryder K, A. Malmborg and E. Söderlind (2017). (Virtual business models: Entrepreneurial Risk and Rewards, https://en.wikipedia.org/wiki/Virtual_business_model

[8] Bryder, Karin, Anki Malmborg-Hager Eskil Söderlind (2016). Virtual Business Models 1st Edition Entrepreneurial Risks and Rewards Write a review eBook ISBN: 9780081001820 Hardcover ISBN:

9780081001417 Imprint: Woodhead Publishing Published Date: 5th February 2016.

[9] Prahalad and Hammel

[10] Mckinsey (2015). "Growing beyond the core business" Report on https://www.mckinsey.com/business-functions/strategy-and-corporate-finance/our-insights/growing-beyond-the-core-business

[11] Lindgren and Bandsholm "The relationship axiom.

[12] Goldmann, Steven L, Roger N. Nagel, Kenneth Preiss (1994). Agile Competitors and Virtual Organizations: Strategies for Enriching the Customer, ISBN: 978-0-471-28650-9

[13] https://www.vocabulary.com/dictionary/virtual

[14] Oxford Dictionary (2018).

[15] Lindgren, P, Yariv Taran, Harry Boer (2010). From single firm to network-based business model innovation International Journal of Entrepreneurship and Innovation Management.

[16] Turban, Efram, David King, Jae Lee, Dennis Viehland [2004] Electronic Commerce – A Managerial Perspective Pearson Isbn 0-13-123015-8

[17] Vervest, Peter H.M., van Heck, E., Preiss, K., Pau, L.-F (2005) Smart Business Network Smart Business Networks Springer Verlag.

Biography

P. Lindgren Ph.D, holds a full Professorship in Multi business model and Technology innovation at Aarhus University – Business development and technology innovation and has researched and worked with network based high speed innovation since 2000. He has been head of Studies for Master in Engineering – Business Development and Technology at Aarhus University from 2014–2016. He is author to several articles and books about business model innovation in networks and Emerging Business Models. He has been researcher at Politechnico di Milano in Italy (2002/03), Stanford University, USA (2010/11), University Tor Vergata, Italy and has in the time period 2007–2010 been the founder and Center Manager of International Center for Innovation www.ici.aau.dk at Aalborg University. He works today as researcher in many different multi business model and technology innovations projects and knowledge networks among others E100 – http://www.entovation.com/kleadmap/, Stanford University project Peace Innovation Lab http://captology.stanford.edu/projects/peace-innovation.html, The Nordic Women in business project – www.womeninbusiness.dk/, The Center for TeleInFrastruktur (CTIF) at Aalborg University www.ctif.aau.dk, EU FP7 project about "multi business model innovation in the clouds" – www.Neffics.eu. He is co-author to several books. He has an entrepreneurial and interdisciplinary approach to research and has initiated several Danish and International research programmes. He is founder of the MBIT lab and is cofounder of CTIF Global Capsule.

His research interests are multi business model and technology innovation in interdisciplinary networks, multi business model typologies, sensing and persuasive business models.

Index